Masterworks of Early 20th-Century Literature
Part I

Professor David Thorburn

THE TEACHING COMPANY ®

PUBLISHED BY:

THE TEACHING COMPANY
4151 Lafayette Center Drive, Suite 100
Chantilly, Virginia 20151-1232
1-800-TEACH-12
Fax—703-378-3819
www.teach12.com

Copyright © The Teaching Company, 2007

Printed in the United States of America

This book is in copyright. All rights reserved.

Without limiting the rights under copyright reserved above,
no part of this publication may be reproduced, stored in
or introduced into a retrieval system, or transmitted,
in any form, or by any means
(electronic, mechanical, photocopying, recording, or otherwise),
without the prior written permission of
The Teaching Company.

Permissions Acknowledgements begin on page 233 and constitute a
continuation of the copyright page.

ISBN 978-1-59803-323-6

David Thorburn, Ph.D.

Professor of Literature, Massachusetts Institute of Technology

David Thorburn is Professor of Literature at MIT and Director of the MIT Communications Forum. He received his A.B. degree from Princeton and his M.A. and Ph.D. from Stanford. He taught in the English Department at Yale for 10 years before joining the MIT Humanities Department in 1976.

Professor Thorburn has won numerous fellowships and awards, including Fulbright, Woodrow Wilson, and Rockefeller Foundation fellowships, and he has lectured widely in the United States and Europe on literature and media.

He is the author of *Conrad's Romanticism* and many essays and reviews on literary, cultural, and media topics in such publications as *Partisan Review*, *Yale Review*, *The New York Times*, and *The American Prospect*, in addition to scholarly journals. Professor Thorburn is also a poet whose work has appeared in *Threepenny Review*, *Slate*, *The Atlantic Monthly* and other publications. He has edited collections of essays on Romanticism and on John Updike, as well as a widely used anthology of fiction, *Initiation: Stories and Novels on Three Themes*. He is the co-editor of *Democracy and New Media* and *Rethinking Media Change*, the launch volumes in the MIT Press series "Media in Transition," of which he is editor in chief.

Professor Thorburn was the founder and for 12 years the Director of the Film and Media Studies program at MIT. He has won teaching awards at both Yale and MIT, where his courses in modern fiction and film are among the most sought after in the Humanities Department. In 2002, he was named a MacVicar Faculty Fellow, MIT's highest teaching award.

Table of Contents
Masterworks of Early 20th-Century Literature
Part I

Professor Biography		i
Course Scope		1
Lecture One	Road Map—Modernism and Moral Ambiguity	3
Lecture Two	How to Read Fiction—Joyce's "An Encounter"	19
Lecture Three	Defining Modernism—Monet's Cathedral	37
Lecture Four	Defining Modernism—Beyond Impressionism	52
Lecture Five	*The Man Who Would Be King*—Imperial Fools	67
Lecture Six	*Heart of Darkness*—Europe's Kurtz	84
Lecture Seven	*Heart of Darkness*—The Drama of the Telling	101
Lecture Eight	*The Shadow-Line*—Unheroic Heroes	116
Lecture Nine	*The Good Soldier*—The Limits of Irony	133
Lecture Ten	*The Good Soldier*—Killed by Kindness	150
Lecture Eleven	Lawrence (and Joyce)—Sex in Modern Fiction	167
Lecture Twelve	"Horse Dealer's Daughter"—A Shimmer Within	182
Timeline		198
Glossary		210
Biographical Notes		215
Bibliography		229

Masterworks of Early 20th-Century Literature

Scope:

This series of 24 lectures examines tradition and innovation in a representative sampling of Modernist fiction—the experimental masterpieces of Joseph Conrad, James Joyce, Virginia Woolf, and others that helped to make the early 20th century a decisive moment in the history of literature.

The lectures assume no prior knowledge of our authors and always begin by offering a basic historical, cultural, or biographical context for the texts studied. They are intended as introductions—a stimulus to read more of these writers—as well as other Modern poets and novelists, some of whom will enter the lectures by way of quotation, reference, and comparison.

Novels suffer far more than poems from the limits imposed by teaching formats. Novels take time to read; they demand or expect a commitment of time and solitary attentiveness. And in the classroom or in a lecture or on the page, we who wish to discuss and analyze novels must inevitably center on fragments, representative moments. No class can do justice to the glorious rush and fullness of a good novel.

These lectures try to minimize this limitation by focusing when possible on novellas—*The Man Who Would Be King*, *Heart of Darkness*, *The Shadow-Line*, *The Metamorphosis*—instead of full-length novels. In a few cases, the lectures focus on stories (D. H. Lawrence, Babel), although in nearly all these instances the chosen texts are compelling works of art and, even in their concision, landmarks of Modernism.

We begin with a modest, practical version of literary theory, illustrating some principles of close reading, and then devote two lectures to the problem of defining Modernism. These lectures, which include discussions of Impressionist and Postimpressionist painting, are also intended to supply a modest context for the literary texts we will study.

Subsequent lectures will center on specific stories or novels, aiming to offer a close reading of their primary themes and formal strategies. We will pay attention to what is distinctive about each of our texts, what marks it as a novel by Conrad or by Woolf. But we will also ask what qualities it shares with other Modernist fiction and how these qualities differentiate Modern novels from their great 19th-century ancestors. The following topics will recur in our discourse: the role of the artist in the Modern period, the representation of sexual and psychological experience, the themes of homelessness and disconnection from received values and belief systems, and the virtues and defects—because there *are* defects and we should acknowledge them—of the aggressively experimental character of so many Modern books.

Lecture One
Road Map—Modernism and Moral Ambiguity

Scope:

This first lecture offers a roadmap of the course—a range of fiction from Kipling and Conrad through Joyce, Virginia Woolf, Isaac Babel, and Nabokov. The archive of commentary that has come to surround most of these authors can be daunting, but this course and these lectures are intended both for new readers and for those returning to these writers after many years. They are grounded in the method of interpretation called "close reading" or "formalism"—terms used to described the principles of the so-called American "New Critics" of the 1940s and 1950s. This approach stresses the primacy of the encounter between the attentive reader and the text itself. This introduction to the course concludes with a partial reading and analysis of John Crowe Ransom's poem "Captain Carpenter," intended to serve as an instance of "close reading" and also as an introduction to the themes and strategies in the fiction we will be reading in the following weeks. The poem's mix of comedy and bizarre violence may be said to generate a distinctively modern experience of moral ambiguity.

Outline

I. This lecture explains the organization of the course and defines some principles of "close reading," stressing their special importance in reading modern fiction.
   A. The lectures are intended as genuine introductions, a stimulus to read more of these writers as well as other Modern poets and novelists.
   B. Novels, as opposed to poems, suffer from the limits imposed by teaching formats.
      1. Novels demand a commitment of time and solitary attentiveness. To discuss them in a course format, we must inevitably focus on fragments and representative moments.

2. This course attempts to minimize such limitations by focusing on novellas and some stories as well as short novels.
C. We will try to minimize the problem of studying extended narratives in another way by stressing the principle of "close reading."
   1. The idea of close reading was the key principle of the so-called New Critics. This work was popularized in the mid-20th century by Cleanth Brooks and Robert Penn Warren.
   2. In this approach, the text itself is primary, not prior information about the authors' lives or historical or cultural information surrounding the era in which the author lived. Also in this approach, close attention to small items, single lines, or single episodes will lead to central insights.
   3. Although background information about the author can be valuable, the reader's encounter and relationship with the text is of primary importance to understanding it.
D. Two scholars, Richard Ellmann and Hugh Kenner, represent two poles of discourse about Modern literature in English.
   1. Kenner's idea of Modernism was more stringent than Ellmann's and, in a certain sense, more elitist and darker. His interest lay in such writers as Beckett and Pound whose vision of the world was despairing.
   2. The difference between the approach of these scholars parallels the difference in tone and world view between Joyce and Pound.
   3. This course attempts to synthesize the two and to include the professor's own perspective.

II. Approaching Modernist writers can be intimidating because of the amount of scholarly material written about them.
   A. If we factor in the widespread notion of the alarming difficulty of Modern literature, it is easy to see why ordinary readers would be discouraged about approaching these writers on their own.
   B. Nonetheless, this course encourages readers to trust themselves and to aim for an authentic, personal connection to our writers.

1. Literary academics have a stake in insisting on the difficulty of books, as this justifies their lives and helps to credentialize them.
2. Immensely valuable scholarship *is* available, of course, and is worth consulting.
3. But whether you are coming to these great writers for the first time or returning to them after a long separation, in this course you should aim for as close to an unmediated connection to the fiction as possible.

III. We will read our authors in historical sequence.
   A. We begin with the Modernists who straddled the 19th and 20th centuries, Kipling and Joseph Conrad; then we read Conrad's collaborator Ford Madox Ford.
   B. The "high Modern" generation of James Joyce, Virginia Woolf, and Frank Kafka are discussed next.
   C. We conclude with a Russian and an American Modernist, Isaac Babel and William Faulkner, and with Vladimir Nabokov, who might be called a "Postmodernist."
   D. This sequence emphasizes a more Anglo-American and "conservative" Modernism instead of the surreal, stringent, and more despairing strain of Modernism we might call "European" Modernism.

IV. We conclude this first lecture by discussing John Crowe Ransom's poem "Captain Carpenter"—an emblem or distillation of many Modernist themes.
   A. On first reading, the poem elicits silence or nervous laughter. Readers are often uncertain about how to answer such questions as, "Do you like this poem? What is it about? Why does it trouble you or other readers?"
   B. Uncertainty and nervousness, it turns out, are appropriate responses to the poem's bizarre comic tone, its ineffectual hero, and its surreal tale of his physical dismemberment.
   C. Close reading will reveal that the poem's language is archaic, invoking the medieval world of chivalry, heroic ballads and Christian values.
      1. The poem has a ballad structure and tone.

2. Captain Carpenter rides into battle like the knights of medieval romance, the Christ-like heroes and saviors of our older literature.
  3. But his comic ineffectualness makes him a mock-heroic figure, something ridiculous and outmoded.
  4. The narrator or speaker of the poem is also part of its meaning. "I would not knock old fellows in the dust," he says. So he, too, like Captain Carpenter himself, is passive and ineffectual.

D. These themes come up repeatedly in Modern literature, which is full of narrators who witness but cannot influence the stories they tell. Some of these narrators will be damaged or unreliable. This split between narrators and actors also suggests a larger theme or preoccupation of Modernism: the separation of thought from action, of intellect and understanding from the power to act.

E. A final paradox of this poem is also a recurring element in many of the texts in our syllabus. Asserting its own modernity, its difference from heroic and religious traditions, the poem nonetheless depends on our recognition of the language and stories of that tradition. That is, its basic meaning and our capacity to understand the poem—and of many modernist texts—are dependent upon the very past the poem aims to judge and transcend.

Permissions Acknowledgment:
"Captain Carpenter", from *Selected Poems* by John Crowe Ransom, copyright © 1924 by Alfred A. Knopf, a division of Random House, Inc. and renewed 1952 by John Crowe Ransom. Used by permission of Alfred A. Knopf, a division of Random House, Inc. For UK/Commonwealth rights, permission granted by Carcanet Press Limited.

Essential Reading:
Ransom, "Captain Carpenter," in *Modern Poems*, Ellmann and O'Clair, eds., pp. 266–267.

Questions to Consider:
1. How does the archaic, religiously inflected vocabulary of "Captain Carpenter" contribute to the tone and meaning of the poem?
2. How should we understand the role of the first-person speaker, who witnesses the protagonist's misadventures and reports them to the reader but takes no action?
3. What is the reader's final judgment of Captain Carpenter? Is he heroic, foolish, or a mingling of both qualities?

Lecture One—Transcript
Road Map—Modernism and Moral Ambiguity

This course examines tradition and innovation in a representative sampling of the fiction of "high Modernism"—the experimental masterpieces of Joseph Conrad, James Joyce, Virginia Woolf, and others that helped to make the early 20th century a decisive moment in the history of literature.

These lectures will assume no prior knowledge of our authors and they're intended as genuine introductions, a stimulus to read these writers as well as other Modern poets and novelists, some of whom will enter the lectures by way of quotation, reference, and comparison.

Novels suffer far more than poems from the limits imposed by teaching formats. Novels take time to read; they demand or expect a commitment of time and solitary attentiveness. And in the classroom or in the lecture or on the page, we who wish to discuss and analyze fiction must inevitably center on fragments, on representative moments. No class can do justice to the glorious rush and fullness of a great novel.

This course tries to minimize this limitation by focusing when possible on short novels, novellas—*The Man Who Would Be King, Heart of Darkness, The Shadow-line,* Kafka's *The Metamorphosis*—instead of full-length novels, and in a few cases I try to focus on stories by D.H. Lawrence and by the strange, powerful Russian writer, Isaac Babel; though in nearly all these instances the chosen texts are compelling works of art and, and even in their concision, landmarks of Modernism.

And I'll try in another way to turn to advantage the limits imposed by the lecture and classroom format. My approach will center repeatedly on specific passages. There is a strategy, a ground principle embedded here that I think is worth making explicit. My teaching theory for these lectures is grounded in what's called formalism, or what used to be called the New Criticism.

In my apprenticeship as a student and young professor at the end of what is sometimes called the Age of Brooks and Warren—Cleanth Brooks and Robert Penn Warren, the great popularizers of the theory of the New Criticism—I learned these principles of literary formalism, and have used them ever since in my teaching. The basic assumption of this formalist dispensation is that the text itself is primary—not prior information about the author's life, not historical or cultural information that may be the possession of specialists or of learned historians—but the text itself and the reader's own authentic encounter with the text is the primary source of literary understanding.

I remember when I was an undergraduate, what a liberating idea this was to a young man who was the first of his family to go to university, who had no special learning or special knowledge, to be told by my professors that if I applied my own attentiveness and intelligence to the text before me, I would understand it was an immensely liberating principle. And I believe that principle is still fundamental for the reading of literature. Don't misunderstand me. I'm not suggesting that learning, knowledge about an author, all kinds of other information is not valuable—it is. But the most fundamental encounter ought to be between the text and the reader's intelligence.

I'll try to develop more systematically these principles of close reading, clarify how they actually should work in practice, in my second lecture, which will focus on the second story in Joyce's collection, *Dubliners*, "An Encounter." And I'll use "An Encounter" as a kind of demonstration piece for an elaboration of these principles of close reading.

There is an ancestry or pedigree for this course that I would like to make known to you, like to make clear. When I began my career as a young assistant professor at Yale, one of the great figures there was the Joyce biographer and scholar, Richard Ellmann. At a certain point Ellmann became ill and I was asked to replace him in the lecture course on Modern fiction that he taught. So at a very early stage in my career I taught through Richard Ellmann's syllabus on Modern fiction; and then later on, after I took over the course on my own and began to teach it on my own, I made certain kinds of revisions and changes to reflect my own perspectives.

Some years later I did a visiting professorship at the University of California in Santa Barbara, and I arrived there one year after the great Modernist scholar Hugh Kenner had left to go, after a 20-year career at Santa Barbara, to conclude his career at Johns Hopkins.

And because my arrival at Santa Barbara coincided with Kenner's departure, I was asked to take on Kenner's Modern fiction course.

Kenner and Ellmann could be said in a certain sense to represent the two different poles of the discourse about Modern literature, especially about Modern literature in English, and teaching through Kenner's syllabus was an immensely instructive experience for me because Kenner's idea of Modernism was more stringent, more austere, in a certain sense more elitist, darker. He tended to be interested in writers who were not only more difficult, but also in writers whose vision of the world was more despairing, more nihilistic.

One way one can understand this difference is to think about the difference between Joyce and Ezra Pound. Kenner was the author, among other things, of a wonderful book called *The Pound Era*. He also wrote many books on Joyce, but *The Pound Era* is his scholarly masterpiece, and in that book he elaborates an idea of Modernism that seems to me in many respects far more elitist than the ideas about literature and the ideas about Modernism that are implicit in the scholarship that we would associate with other figures, and especially with Richard Ellmann. So we can think of the conflict between Joyce and Pound or between Kenner and Ellmann as a kind of embodiment of the poles of discourse that have organized discussions of Modern literature since they did their work.

What I'm offering is a kind of synthesis of that work with a good deal of overlay now, after many years of my own teaching and scholarship, of my own perspective on these matters. I won't hide the fact that I'm much more of the school of Ellmann than of the school of Kenner, but it's important to recognize that Kenner himself, although he never acknowledged this, moved more and more fully, especially in his writings on Joyce, toward a more humanistic and a more generous and a more democratic idea of the meaning of Joyce's work than that with which he began. In any case, these two defining scholarly positions inform the course I'm about to offer in these lectures.

I have one other observation to make, or comment to make, about the assumptions we should be making as we approach these difficult and challenging writers, and the subtitle for this aspect of my argument might be "Against the Scholars (Not Exactly)". Maybe a better subtitle would be "Against (Professional) Interpretation," because I want to acknowledge at the outset that any rational person thinking about approaching these writers has of course got to be intimidated and disturbed by the mountain of scholarship that surrounds all of these figures.

There are libraries, archives of material that have now been generated on every one of the writers that we'll be talking about in this course. A collection of all books in English—just in the English language, not to mention an immense trove of other languages, in both eastern and western European languages—on a single one of many of our writers would fill not a room but a building, a library.

I recently received an advertisement from a single American university press, which offered seven new scholarly books just devoted to Joyce. Now, think about what that means: the number of scholarly presses there are in the world—and in fact seven books from many scholarly presses is close to the total number they might offer in a year—and this press was offering seven separate titles on Joyce. The number of books and articles that have been written on figures like Joyce and Virginia Woolf, numbers not in the thousands but in the tens of thousands; of course this is an intimidating avalanche of material.

And then when we factor in the very widespread and partly accurate notion of the alarming difficulty—even, some people would argue, the impenetrability—of Modern literature, it's easy to see why ordinary readers would be discouraged about approaching these giants of Modernism, and especially approaching them on their own without help. Nonetheless, that is what I am recommending.

If you want an authentic connection to a writer, I tell my students, spend your time with the texts on the syllabus, or with other works by the same author. Leave the professional criticism in its specialist, credentializing world. There's a good reason for this; partly because it's important to try to establish an authentic personal relation to the material you look at, but it's also because there's a kind of anthropology of literary scholarship that we need to be aware of. We might almost say that literary study has become a kind of mass production or industrial activity.

There's a professional elite in the United States especially, although it's spread now to England and parts of Europe as well, which has a stake in making books look difficult because they want their services to be required. Many interpreters of Modern literature act like priestly intercessors who try to stand between the text and the reader, and say, "You ignorant readers need me in order to understand the text." They have a professional incitement or encouragement to do that. And of course there's also, and this is the worst aspect of the professionalization of literary study, since promotion and tenure depend upon what is called "literary productivity," there's an immense incitement for younger scholars especially to create only the most nuanced differences between themselves and other interpreters in order to justify the production of more interpretations and the production of more scholarly studies.

Now, again, don't misunderstand what I'm saying. I don't mean that there isn't also an immensely valuable trove of scholarship devoted to all of these writers. There is. And part of my obligation as a teacher is to serve as a kind of conduit for the best of those ideas. I try to do that in my teaching in university, and I'm going to try to do that in these lectures as well.

Nonetheless it seems immensely important, especially when we approach Modern literature, to trust ourselves and to trust the texts, and that's what I'm urging. Whether you're coming to these great writers for the first time, or returning to them after a long separation, I urge you to aim for as close to an unmediated connection to this fiction as possible. My own discourse will always situate itself, such that it should be explicable even if you've not yet read these writers. And if you have read them in the past and haven't yet got back to them, I hope that my commentary will be an encouragement to return to the writers with renewed energy and renewed understanding.

Let me offer a very quick tour of where we're going in the course. We begin in the 19th century with a transitional figure, Rudyard Kipling, who has one foot in the 19th and one foot in the 20th century. Thence to Joseph Conrad, the first of whose texts we read was written in 1898, *Heart of Darkness*; the second great novella of his that we're reading was written in the 20th century. And then we pay attention to other figures I think of as first generation Modernists beyond Conrad. We read also a novel by Ford Madox Ford published in 1915.

Then we move on to the great figures whose most important work appeared in the 1920s—to Joyce and Woolf and Kafka and Isaac Babel. Thence to Faulkner, who did much of his great writing in the

'20s, but the book we're reading of his, *Absalom, Absalom*, one of his great masterpieces, appears in the '30s; and then finally as a kind of coda, our sort of Postmodern Modernist, Vladimir Nabokov.

That's a kind of quick tour. This sequence lays emphasis on a more Anglo-American and "conservative" Modernism than on the surreal, stringent, more despairing strain of Modernism we might call European. I'll return to this distinction in the third and fourth lectures, which will aim to situate literary Modernism within the larger culture of late 19th and earlier 20th century thought.

But I hate first classes that are all generalities and technical matters—how many assignments? What are the due dates?—that are all map and no taste of the local cuisine. So I'd like to conclude this first lecture with a distillation of many of the central themes of the adventure before us. And my way of doing that is to focus on a famous American poem by the American poet John Crowe Ransom, titled "Captain Carpenter," written in 1924, right in the middle of the great decade that produced Joyce's *Ulysses*, T.S. Eliot's *The Wasteland*, Ford Madox Ford's tetralogy of novels *Parade's End*, Virginia Woolf's *Mrs. Dalloway* and *To the Lighthouse*, and many other masterpieces of Modernism.

I'm going to read a good part of this poem to you. It's 16 quatrains (16 stanzas). I'm going to cut out a few of them but I want you to feel the whole poem. When I do this in my classes I read the entire poem, and you'll get a flavor for it; you'll hear most of it.

Captain Carpenter

> Captain Carpenter rose up in his prime
> Put on his pistols and went riding out
> But had got well-nigh nowhere at that time
> Till he fell in with ladies in a rout.
>
> It was a pretty lady with all her train
> That played with him so sweetly but before
> An hour she'd taken a sword with all her main
> And twined him of his nose for evermore.
>
> Captain Carpenter mounted up one day
> And rode straightway into a stranger rogue
> That looked unchristian but be that as may
> The Captain did not wait upon prologue.

But drew upon him out of his great heart
The other swung against him with a club
And cracked his two legs at the shinny part
And let him roll and stick like any tub.

Captain Carpenter rode many a time.
From male and female took he sundry harms.
He met the wife of Satan crying "I'm
The she-wolf bids you shall bear no more arms."

Their strokes and counters whistled in the wind
I wish he had delivered half his blows
But where she should have made off like a hind
The bitch bit off his arms at the elbows.

And Captain Carpenter parted with his ears
To a black devil that used him in this wise
O Jesus ere his threescore and ten years
Another had plucked out his sweet blue eyes.

Four stanzas I'm skipping.

I would not knock old fellows in the dust
But there lay Captain Carpenter on his back
His weapons were the old heart in his bust
And a blade shook between rotten teeth alack.

The rogue in scarlet and grey soon knew his mind.
He wished to get his trophy and depart
With gentle apology and touch refined
He pierced him and produced the Captain's heart.

God's mercy rest on Captain Carpenter now
I thought him sirs an honest gentleman
Citizen husband soldier and scholar enow
Let jangling kites eat of him if they can.

But God's deep curses follow after those
That shore him of his goodly nose and ears
His legs and strong arms at the two elbows
And eyes that had not watered seventy years.

The curse of hell upon that sleek upstart
That got the Captain finally on his back
And took the red red vitals of his heart

And made the kites to whet their beaks clack clack.

When I read this to my classes there's silence, of course, nervous laughter, coughing. A reluctance to respond when I say, Do you like this? What is it about? Why is it troubling you?

But after a while I get the class to begin to talk, and of course one of the things I do is try to make them see that their uneasiness, that their uncertainty is intended as part of the response; that the bafflement and uncertainty that they feel about how to respond to this material was certainly an intended effect of the poem. And I say, What is it about the poem that makes us feel so uneasy, that makes us uncertain about how to respond?

And the first thing we say is that it has a kind of bizarre comic tone. What is this dismemberment plot that's acted out here? How are we supposed to respond to it? Does the poem respect and admire Captain Carpenter or does it think he's some kind of a ridiculous fool who's a kind of clownish joke? And I don't quite answer the question, but it's obvious that the poem creates this uncertainty about how to read him, how to understand him.

Then we look at certain features of the language, and I make the students listen especially to the archaic tone of the language, to the fact that the diction is full of religious and chivalric terms. "God's mercy rest on Captain Carpenter now." "He met the wife of Satan." All kinds of religious references. And the idea that he rides out to do battle as if he's on horseback and is some sort of a chivalric knight.

And our discussion of these matters then leads further to the recognition that the quatrains and the rhyme scheme to which the poem is committed has a kind of flavor of a ballad. And once we've made these connections, the connection between the archaic, partly religious chivalric diction, and the connection to the ballad structure, it's an easy next step to make the students recognize that the poem has a kind of mock heroic character, that it seems to invoke the kind of narrative ballad that has been characteristic of certain forms of English and American poetry and certain forms of European poetry as well since at least the Middle Ages; and that there are many poems and even songs that use this kind of ballad structure and that often tell about some hero, some admirable figure.

In the Middle Ages, in England especially, there are certain poems that are literally about knights in shining armor who ride out to do battle with evil and to rescue damsels. And what we come to recognize is that the poem invokes and plays against this expectation; that its language and its form are like those older poems, but that the content of the poem is not like that, that the content of the poem has this bizarre dismemberment tale in it.

And the argument then is that the religious literary overtones invoke the quest heroes, the knights in shining armor, the Christ-like saviors of earlier stories and earlier cultures. In fact the idea of associating the knight who rides out to do battle with Christ himself was an explicit one in the Middle Ages and after, and these figures were often understood as wayfaring war-faring Christians who were types of Jesus Christ, who were aggressively carrying Jesus' power into the world and protecting the world from evil, rescuing people from evil.

The poem invokes all those values, but of course this wayfaring warfaring Christian is a character so ineffectual that his nose is cut off, his ears are cut off, he rolls and sticks like any tub. What are we supposed to think about this? And the answer, of course, is we are supposed to respond with a kind of ambivalent, bizarre uncertainty with which we began; but now we no longer have to feel uneasy about it or feel that we're misunderstanding what's happening, because we recognize that it's at the heart of what the poem intends us to say. It's as if the poem is calling into the question the status of those ethical norms that are embodied in those older texts: not just Christ himself, but citizens, soldiers, scholars; all are rendered impotent, ridiculous, problematic, outmoded by the bizarre dismemberment plot that the poem enacts. By the end of the poem, Captain Carpenter rolls and sticks like any tub.

Then there's a further aspect of the poem that I call students' attention to, and in some classes the very best students recognize this and ask about it before it's necessary for me to encourage them to find the passage, and that has to do with the strange way in which at a certain moment in the poem, in the sixth stanza, we hear this line: "I wish he had delivered half his blows." There's a first person narrator, a first person witness in this poem, and he's amazingly, disturbingly passive. He's clearly on Captain Carpenter's side. He's unambiguous in his sympathy for Captain Carpenter, but he's also immensely passive. He says, "I would not knock old fellows in the dust"—but there lay Captain Carpenter on his back. He watches it happen.

This dramatizes something that comes up again and again in Modern literature, and we'll see versions of this later on in the course repeatedly—the separation of thought from action, of understanding from will. It's a recurring dichotomy in Modern fiction, which is full of narrators who are impotent to change the events they try to understand and recount to us. You can feel the narrator's impotence in the final stanza especially. "The curse of hell upon the sleek upstart / That got the captain finally on his back / And took the red red vitals of his heart /And made the kites to whet their beaks clack clack."

I remember the first time I read the poem when I heard the "clack clack," I knew I was in the presence of heart. The way the poem leaves language and makes us feel the power and the danger of those carrion birds. But the most important point about the ending of the poem is the way it emphasizes the astonishing passivity of the narrator, the impotence of the narrator. Again and again in Modern fiction we are going to encounter impotent narrators who often are afflicted by an understanding of what they're seeing but are unable to act. We also are going to encounter some narrators who are unreliable, and whose accounts of what they tell us we need to question. But in either case, whether we're talking about unreliable narrators or narrators who understand deeply what's going on, what we also encounter essentially is the principle of the passive witness rendered incapable of action by their attempt to understand: a separation of thought and action, of intellect and the power to act.

And there's a final paradox in this poem that I also think is important to call attention to and that I want my students to think about. It's one of the deep paradoxes of Modernism, and virtually every single Modernist text implicitly contains this contradictory or complex energy. And this paradox has to do with the fact that there's a certain way in which the poem very aggressively insists on its newness, on its Modernity.

It invokes all those old values only to say they don't work any more. In other words, the poem is about how in Modernity the old dispensations, the old inherited values, the old categories for making sense of the world and for judging people no longer hold; that the modern world is a more problematic and a more dangerous and maybe a much more morally ambiguous place than our older time was.

And there's a kind of, if not arrogance, there's a kind of aggressiveness in this assertion, and it's at the very heart of our understanding of the poem; and yet there's a paradox here. The paradox has to do with the fact that the poem, in order to make its argument about its newness, in order to make its argument about its Modernity, the poem must rely on the very strategies, the very values, even the very literary forms that it is claiming to overgo. There's a tremendous irony, a great paradox here, and this is a paradox that's at the center of a great deal of Modern literature. It's played out in different registers in Thomas Mann's great novel *Joseph and His Brothers*, in T.S. Eliot's *The Wasteland*, in Faulkner's *Absalom, Absalom*. It's played out with immense complexity and richness in James Joyce's *Ulysses*. It's a paradox we'll return to and savor as we go on in the course. It's the paradox that these Modernist texts, in asserting their newness, depend upon the very past they pretend to transcend, depend even upon the very categories and the very assumptions about value that they are also claiming are outmoded. This is a paradox, a contradiction, a complexity to which we will return again and again in the lectures that follow.

Lecture Two
How to Read Fiction—Joyce's "An Encounter"

Scope:

This "empowerment" lecture is intended to clarify the principles on which the reading of most literature depends, especially the principles we must embrace to read the challenging Modernist texts of this course. Contemporary literary scholarship can seem daunting and can encourage a sense of literature as a code that must be broken, a code that only the learned can understand. Using James Joyce's short story "An Encounter" as our case in point, we'll show how attentiveness and common sense are the most essential attributes of good readers.

Outline

I. Understanding most literature—especially Modernist texts—depends upon the individual reader trusting to his or her own close reading as opposed to relying on literary criticism for insights.
  A. Stories should always be read with our first principle in mind: what the task of reading is and the nature of literature.
  B. Despite the remarkable scholarship in literary study, one of its consequences has been to encourage a very narrow and, in some ways, dangerous idea of what literary study is.
    1. The danger might be summarized as the "puzzle-solving school" of literary study, which asserts that the language of literature is out to trick readers and open only to specially tutored decoders.
    2. This view transforms the act of reading into a game in which the reader is manipulated like a mouse in a maze. It also transforms writers into mere conjurers whose art is that of riddle-making and trivial deception.
    3. Further, this attitude undermines what is most important about literature because it assumes that readers must not accept what most literary works seem to be saying directly and clearly and that their true meanings lie

concealed beneath an ocean of details whose function is to mislead readers.

C. But the far happier truth is otherwise. Nearly all literature speaks to most literate adults, not to professors or other specialists. In fact, literature is one of the few activities left in our era of specialization that does not require expertise or specialized knowledge.

II. To demonstrate the importance of these principles, and to illustrate something of the nature of "close reading," let's examine James Joyce's short story, "An Encounter."

A. The story relates the modest adventures of two friends playing hooky from their Catholic grammar school. Their plans for an exciting day go awry and in the story ends in their encounter with a strange old man.

B. This straightforward story has great complexity.

C. But for many people, what is straightforward seems uninteresting or simple. To illustrate this in my classes at MIT, and to encourage students to be skeptical of interpretations that feel outlandish, I read excerpts from William York Tindall, *A Reader's Guide to James Joyce* (1959).

1. Tindall discovers a search for the Trinity in the story (in a story that barely mentions religion) by twisting or appropriating certain minor phrases. But I read his most outlandish passages with a straight face and then ask my students what they think.

2. Some embrace Tindall's interpretation and try to extend it, while others admit that they don't recognize the story they read in his comments.

3. At some point, I slam the book to the floor, declaring Tindall's interpretation "Garbage." The mere publication of an interpretation does not guarantee that it will be sensible, much less persuasive and illuminating.

4. Tindall's "errors" include:
- indifference to the manifest content—the story, characters, setting
- dependence on "secondary evidence"—such as biographical facts, references to other

works by the same author, information drawn from literary or cultural history—for every central claim in his interpretation. Secondary evidence is valuable, can and should be used, but should never be used as the primary or sole evidence for the meaning of a text.
- Tindall's deepest error is underestimating the complexity of ordinary experience and the importance that Joyce and his story attribute to it.

III. How might we establish a more compelling interpretation of "An Encounter" and, by implication, most literary works?
   A. We would begin with the simple act of description, looking closely at the story and describing its fundamental elements: characters, plot, and language.
   B. Description is a kind of interpretation, allowing for a commonsense approach to the reading of literary texts. As one great teacher (Irving Howe) once said to me: *"Never ignore the fascinating surface in favor of the mere depths."* So any good interpretation of a story must account for its fundamental features: its plot, characters, and its style or language.
   C. The power of "An Encounter" comes from its surface elements. Broadly, the story is a drama about thwarted expectations.
      1. Often in the story, plans go awry or expectations are unmet.
      2. These smaller events reinforce the larger experience of the main action of the story. The plot, then, is a drama of disappointed hopes.
   D. The characters, Mahony and the narrator, show fundamental differences. As we can see by looking closely at their behavior at various moments in the story, the narrator clearly represents imagination and intelligence. He is smarter than Mahony but also timid, and his dumber friend is the one who acts in the face of difficulty.

- E. The difference in character is strongly illustrated when Mahony runs off across the field, completely unintimidated by the old man, while the narrator remains next to him.
    1. The narrator's bragging to the old man suggests that he identifies with adults more than with children.
    2. The narrator gets his comeuppance in a certain moral sense when he recognizes that there is something strange and terrible about the old man, while Mahony runs in freedom across the field.
- F. "An Encounter," then, is partly about the limitations of intellect and intellectuals. Although Joyce himself was intelligent and learned, he frequently dramatizes what the narrator of this story acknowledges in the end: our dependence on ordinary humanity.
- G. The symbolism in "An Encounter" is complex but accessible and reinforces the themes already discussed.
    1. First, what are symbols? They are items in the physical universe that are granted larger meanings not usually associated with them.
    2. How can we tell when an item in fiction is symbolic? There are so many objects in stories that we could not make sense of them if every element were symbolic.
    3. But the fact is, the text will signal what is symbolic; there will be clear encouragements from the story to read certain items symbolically.
    4. In "An Encounter," the green eyes are symbolic, and their symbolic role is obvious, even heavy-handed. The narrator searches among the sailors for "green eyes," clearly a symbol of adventure and freedom. Then he encounters the "bottle green eyes" of the old man. Thus, the symbolism in the story reinforces and enlarges what we've already seen is dramatized in the plot and embedded in the characters.
- H. The style or language of Joyce's story also makes a crucial contribution to our understanding.
    1. Joyce establishes a delicate balance of sympathy and judgment for his narrator in the subtle language of the story. The first-person perspective establishes a basic

sympathy but the diction itself sounds a note of gentle mockery and judgment. This is the vocabulary of an older man, looking back at a younger, naïve self.

2. This rich style, creating both immediacy and moral distance, keeps the story from being grandiose. This speaker knows it takes many lessons to make a man, not just one encounter with a pathetic old pervert.

I. The complexity of Joyce's voice in "An Encounter" also reminds us that this narrator's good sense and humanity are also essential qualities for readers of literature and far more valuable than a specialist's trained eye.

Essential Reading:

Joyce, "An Encounter," in *Dubliners*.

Supplementary Reading:

Thorburn, "Introduction," *Initiation: Stories and Novels on Three Themes*, pp. 1–10.

Questions to Consider:

1. Why does the narrator think it important to tell us that the old man had green eyes? Is there another, larger discovery implicit in this one?

2. Why does the protagonist feel penitent about his negative feelings toward Mahony, and why is this revelation disclosed in the final sentence of the story?

Lecture Two—Transcript
How to Read Fiction—Joyce's "An Encounter"

I think of this lecture as my empowerment lecture, intended to clarify the ground principles on which I believe the reading of most literature must depend, and especially the ground principles we must embrace if we're going to read the challenging Modernist texts on our syllabus.

Now, ideally, we should always begin the process of reading a story or a novel or a poem with our first principles in mind, with a general conviction about what the task of reading is, and about the nature of literature. But this is a very difficult assignment, I think, in our era of the expert. I mentioned in the earlier lecture that the professionalization of literary study after the Second World War had some pernicious consequences.

I don't mean to imply that it didn't have also wonderful consequences. There is a remarkable scholarship in literary study that's a function of this professionalization, but it did have negative consequences as well. And one of those consequences has been to encourage a very narrow and in some ways, I think, a deeply dangerous idea of what literary study in general is, and especially a dangerous idea—somewhat encouraged by the texts, I must admit—of what Modernist literature is like.

This dangerous tendency might be summarized in this way. People who believe in that, or who are affected by these tendencies, are members, one might say, of what I would call the puzzle-solving school of literary study. According to this school, the language of literature is treacherous, cunning, out to trick us, is open only to specially tutored decoders, experts in the hieroglyph, masters of the labyrinth. In this view of literary study it seems to me, at least in its extreme forms, the reader is reduced to something like a mouse in a behaviorist maze, who is rewarded for his cleverness in not allowing himself to be tricked.

Apparently a respectful attitude toward literature in an era that so emphasizes specialization, it actually of course is an attitude that undermines what's most important about literature; because it's surely a very curious respect which assumes that one must guard against accepting what most literary works seem to be saying directly and clearly, that assumes that their true meanings or their hidden meanings lie concealed beneath an ocean of details whose function is to mislead uncautious or inexperienced readers.

I think as I've suggested that this view of literature transforms the act of reading into a kind of game in which the reader becomes a kind of manipulated creature like a mouse in a maze rewarded for its cleverness; and I think it also transforms writers into mere conjurers whose art is that of riddle-making and trivial deception.

What offends me most, I think, about this view of literature is the way it goes against what for me is one of the prime claims literature has to make on human beings, and it's one of the reasons I became a teacher. It's one of the reasons I remain a satisfied teacher, a teacher who gets pleasure from his work; and that has to do with the idea, also implied in something I said in the first lecture, that literature is available to everyone. It's one of the very few activities left to us, in fact, in our era of specialization, which does not require specialized understanding; which wants and intends to be understood by physicists, by biologists, by all manner and form of nerdly MIT undergraduates, by mechanics, by doctors, by laborers.

There's something fundamentally amateur about literary study, and many professors find this troubling, but I find it inspiring because, of course, the root of the word amateur is *amor*, the Latin word for love. You do things for love when you're an amateur, not because you're being paid to do it or because it's necessary to your livelihood. And I think in fact that one of the great glories of literature is that it is still possible to see it in these ways, even, as I hope this course will demonstrate, Modernist literature.

Now, one way I try to demonstrate the plausibility, the power, of these principles of common sense in my classes is by asking my students to read the second short story in Joyce's collection of stories, *Dubliners*, "An Encounter." Let me remind you about what "An Encounter" is. It's a very simple story on its face and in fact, in its substance as well, despite what some scholars have said about it. It's only 9 or 10 pages long in most editions, well under 4,000 words. The experience of reading it even for a slow reader can't take more than 15 or 20 minutes.

In the story an unnamed narrator, fairly obviously the same person, the same voice we hear in the first and third stories of *Dubliners*, tells of his modest adventures on a day in which he and a comrade play hooky from their Catholic grammar school. They don't attend school one day. Their adventure is intended to reach a place called the Pigeon House, which is Dublin's electric station on the breakwater in the bay. They never get to the Pigeon House and instead they have an encounter with a strange old man who by implication is some kind of a pervert, some kind of a pederast, although he does nothing, he just is a scary character. So the encounter is with strangeness, maybe with perversity. It's a relatively straightforward story although in its straightforwardness there is, I think, great complexity.

Now for more than 35 years I've used this story to play a kind of trick on my students. After they've read the story I begin by reading excerpts from a book about Joyce by William York Tindall, a book published in 1959. It's called *A Reader's Guide to James Joyce.* And let me now read a quick excerpt from this study of Joyce:

> The present quest is for the Pigeon House, Dublin's electric light and power station on the breakwater in the bay. Light and power suggest God, and the traditional icon of the Holy Ghost is the pigeon, [Tindall writes] as we could infer from the first and third chapters of *Ulysses*, where pigeon and Pigeon House reappear. The quest, therefore, can be taken as a search for the third member of the Trinity; or since Father, Son, and Holy Ghost are one, as that hunt for the Father, which was to become a theme of *Ulysses*.

Now I'm going to skip a little bit and read another quick passage from this same book:

> The queer old man whom the boys encounter near the bank of the river at Ring's End resembles Father Flynn in clothing, teeth, perversity, [Father Flynn is a character in the first story of *Dubliners*] and preoccupation with ritual. Obsessed with hair and whips, his mind as if unfolding some "elaborate mystery" moves slowly round and round like the goats of Stephen Daedalus' nightmare vision in Joyce's novel *A Portrait of the Artist as a Young Man*.
>
> Never having read Krafft Ebing and not yet aware of what he has encountered, the boy is uneasy at first and at last frightened. The pervert's one action is so extraordinary that even insensitive Mahony, the boy's companion, calls the

ritualist a queer old josser. Nothing in Joyce is unconsidered or accidental. Josser, according to Webster, can be English slang for a simpleton, which hardly seems to apply, but the word can also be pidgin English for a deviltry of a joss or god. Compare the phrase "Lord Joss" from *Finnegans Wake*.

Probably not God as some have thought, the pervert may imply what men unable to reach the Pigeon House find in place of him. *Pigeon*, a suppressed pun, may be to *pidgin* as perversion or defect of love to love itself. In this sense the old josser with his ritual and his desire to imitate others into the mystery could suggest the church, burner of Irish joss sticks.

There's much more in this vein, but I'll stop now. When I read this to my students, it's a kind of trick. In my younger days I used to let it go on for a long time before I reveal that I don't really approve of such interpretations. As I've gotten older and less cruel as a teacher, I usually stop it after a relatively short time. But I usually begin by saying: What did you think of this? How did you like it? And one irony is that it's often the better students, the ones who are most eager to impress the professor, who immediately embrace the interpretation and try to extend it in certain ways.

But eventually some student, however tentatively, will venture the idea, perhaps in an embarrassed way or a uneasy way, will say that her reading did not recognize the religious dimension at all, that somehow she just didn't—maybe there's something wrong with her—but she really didn't get it. And when that reservation is expressed, no matter how tentatively, I take that as my cue to say, "This is garbage!" and I throw the book on the floor, or throw it against the wall.

In its most dramatic history I threw the book, intending to throw it against the wall, when I was teaching at Yale, and it went out the window of the second floor of Connecticut Hall and landed at the feet of someone who was walking into his office at the time. It was the president of Yale University, Kingman Brewster, who picked up the book and brought it back and handed it to me while the class was going on. The terribly damaged condition of the book is a consequence not of the fact that it's been so heavily read, but because it's hit so many walls in the course of 35 or so years.

Now, of course, in a certain sense this is merely a demonstration, a kind of trick, but I think it's a useful one because it encourages the students to recognize that just because a thing is in print doesn't mean that one needs to believe it abjectly. It encourages a lack of reverence, a kind of skepticism toward the kinds of things that literary critics say about texts.

And then what I do is I ask students to talk a little bit about what it is in Tindall's interpretation that seems problematic. It's almost always the case that as soon as their own unease about such interpretation is ratified that the students will begin to raise very strenuous objections to what has been said in that interpretation. It seems to me a perfect example of what I called, before, the puzzle-solving school of literature.

It's almost as if, impatient with the surface elements of the story, Tindall wants to find a deeper or more complex meaning; and I don't want to spend too much time on his errors, on the ways in which Tindall is mistaken, but it does seem useful, I think, to spend a little bit of time on what's wrong with the way he approaches the text.

One of the most dramatic ways in which Tindall's interpretation is ungrounded is that it's based almost entirely on what I'll call secondary evidence. Now secondary evidence of all kinds is valuable to understand literary works, and there's no reason why facts about an author's biography, or materials drawn from other things that the author has written, or ideas that one has drawn from the history or context in which the author lived or in which the story is set, all of that material is legitimate fodder for interpretation. But one can't found one's interpretation on these matters.

One may use these secondary sources, these secondary forms of evidence, to inform one or to alert one to what might be in the text, but in the end your interpretation must be grounded on what's inside the text. Sometimes what I do when I try to illustrate this principle is I'll draw a rectangle on the blackboard, and I'll color the rectangle in with the side of the chalk, and then I'll take a very tiny little corner, the very upper left-hand corner, a tiny little part, and I'll say, okay, Tindall's interpretation is based on this—one line in the text—but all of the rest of this—the plot, the character, the style in which the story is written, 99 percent of the words in which the story is written— none of that has anything to do with the interpretation. If the interpretation is grounded on this tiny little corner of the text, it can't be persuasive. It can't be plausible. To be plausible an interpretation has to be grounded in what is manifest, central, fundamental to the story.

Another aspect of Tindall's interpretation, which I can't resist at least briefly mentioning, is the way in which his discourse generates its own evidence. This is, perhaps, the most contemptible aspect of his interpretation. Remember what he says: "The boys are going to the Pigeon House." So then he discovers in the story something he calls pidgin English, and he calls this a suppressed pun. But that has nothing to do with anything in the story. Joyce certainly never had any intention; the story has no intention, or no capacity, to remind us of such a thing.

And the particular detail that he cites: he uses the word joss, and he says the boy says, "Oh look. There's a queer old josser," says Mahony. And then Tindall looks at this line and says, "Well, I'll look in the dictionary." The dictionary provides him with a definition, and he says, "The dictionary says that josser might mean simpleton, which hardly seems to apply." Well why does it hardly seem to apply? Why couldn't the boy be saying, "Look at that queer old simpleton"? "Look at that strange old jerk there"?

Of course in some ways it's the most *obvious* interpretation of what the boy is saying, but Tindall wants to find a much deeper symbolic meaning, and he finds that symbolic meaning not by looking in the text itself, not by looking in the story "An Encounter," but by drawing on a line that he finds in Joyce's last, most difficult, most unreadable novel, *Finnegans Wake*. Well, if you try to interpret "An Encounter" on the basis of evidence that you found in *Finnegans Wake*, you're interpreting an invention of your own. You're not really reading the story.

There are other egregious errors, violations of common sense in Tindall's interpretation, but let me end by simply saying that one of the things he seems to reveal is an impatience with what might be called the manifest content of the story. Because the story seems to be about boys who take a day off from school and play hooky and have an encounter with strangeness, Tindall implicitly seems to feel that this is trivial in comparison to something important, like a story about religious themes, a story about the search for the Trinity. So he simply substitutes an interpretation that he finds more powerful or more important.

As I hope to demonstrate before we're through with this lecture, Tindall's deepest error is to underestimate the complexity of ordinary experience and the importance that Joyce himself and his story attribute to that ordinary experience.

Well, how might we go at a more powerful, a more compelling interpretation of "An Encounter," and by implication all literary works, or most literary works? My answer is relatively straightforward. We would begin with the simple act of description. We would require ourselves or our students to look closely at the story and to describe its fundamental elements as carefully as possible.

And I sometimes will say to my students things like: What are the most fundamental constituents of a work of fiction? They'll stumble around a bit, but most of them are pretty quick to come to the conclusion, come to the recognition, especially if I encourage them, to say the most obvious things. I encourage them to say this because I believe, with a great teacher of my own who said this to me almost 50 years ago, he said, "Never ignore the fascinating surface in favor of the mere depths." I love the line. "Never ignore the fascinating surface in favor of the mere depths." There's a notion that depths are great and surfaces trivial, but that's never true of literary works. Incidentally, I think it's never true of life either, but that's another matter.

So we would begin with the act of description, and we would try to describe the most fundamental constituent elements that all fiction has—characters, plot, and language. All stories are written in language. All stories have some kind of character; even if there are animals in the story, the animals are the characters. And all stories have some kind of momentum or action.

So in our description of a literary work, our most fundamental obligation at first is to describe those constituent elements as carefully and as effectively as we can. One of the reasons I like the word description is that it demystifies the act of interpretation, because what it suggests, anybody can describe. It doesn't require special knowledge. You don't have to have gone to graduate school in order to learn how to describe something, in order to look at something and describe it carefully.

And once you realize that description is a kind of interpretation, that description leads to or is the ground of all interpretation, I think you're liberated into a kind of commonsense approach to the reading of literary texts.

And if we follow these principles with regard to "An Encounter," what we would find is not that the story was trivial or small, but that it was deep and powerful, but its depth, its power comes from its surface elements. We would describe, for example, its plot, and what we would discover (I've already described the plot in its broadest outlines) is not only does the broadest outline of the story enact a basic drama in which one's expectations are thwarted by experience—the boys go in search of the Pigeon House and encounter a pervert; that's a description of the larger action of the story—but what one would also discover is that many of the smaller passages or details in the story also enact a similar pattern. In the very first paragraph of the story, for example, we're told that one of their playmates, who plays very wildly and raucously in his own way, everyone is surprised to learn has a vocation for the priesthood. And so there's an example of an unexpected outcome, of a surprising outcome.

And in fact again and again in the story what we find is that certain plans go awry, that certain expectations are thwarted. Not just that the boys expect to get to the Pigeon House and meet a strange old man and the boys never get where they're going, but also for example earlier in the story the boys plan their day of adventure where three of them are supposed to go, but when the day comes one of the boys doesn't show up. He sort of chickened out. He's cowardly.

And the narrator says, "Well, what should we do?" And his friend Mahony says, "Forget it. We'll go without him." So there's another kind of thwarted expectation. And there are actually dozens of examples of this in the story, in which the largest plot is reinforced by what we might call a series of smaller plots, smaller events that reinforce the larger experience of the main action of the story.

If we move on to character and we begin to describe the characters closely, one of the things we begin to realize is that there's a link between plot and character, and that if we describe the plot carefully we will also begin to get a grip on the nature of character. I mentioned a moment ago that moment when the two boys get together just before they're ready to start their adventure and the narrator discovers that the third friend hasn't come, and Mahony says, "Forget it. Let's go on our own," that we'll go without him.

And it turns out that again and again in the story, when they encounter difficulties or surprises, it's Mahony, not the narrator, who takes action. It's the narrator who planned the day, it's the narrator who had the imagination to be unhappy with their conditions at school, but it's Mahony in the face of actuality who has the spontaneous energy to act. And once we pick up on that detail, we begin to realize that the story systematically articulates a fundamental difference between the narrator and his friend Mahony. The narrator clearly represents imagination, intelligence of a certain kind. He's smarter than his friend Mahony. He thinks ahead. He plans. He's able to conceive adventures. But it's Mahony who can act in the face of difficulty.

And, of course, this difference between Mahony and the narrator is articulated or dramatized dozens or many times in the story. One minor example: When they begin their adventures they come across a bunch of Protestant school kids (they come from a Catholic school). That tells us something about the social and political context within which Joyce's fiction always operates. And Mahony proposes that they charge them and throw rocks at them, but the narrator says, "I protested that the boys were too small."

But by this time in the story, if we've been noticing the differences between the narrator and Mahony, we might be able to draw a different conclusion from the narrator's remarks. Maybe he's reluctant because he's not as eager to fight, because he's not as spontaneous, because the easy activeness of Mahony is something that's much harder for the narrator.

And, of course, this difference between the narrator and Mahony reaches a kind of terrifying climax in the conclusion of the story, where Mahony and the young boy encounter this strange old man, and Mahony runs off across the field completely unintimidated by the old man, and the narrator remains next to him; and the strange old man recognizes in some sense that the narrator has more in common with him than Mahony.

The narrator even brags to the old man that he reads more than Mahony, and it's almost as if, like many intelligent young kids, the narrator of the story, probably the young Joyce in some basic way, is precocious, identifies with adults more than with children, is eager not to be associated with a boy that he thinks is more ignorant than he. And, of course, he gets his comeuppance in a certain moral sense because when he recognizes that there's something strange and terrible about the old man, it's as if he's frozen in fear looking into the face of corruption, while Mahony runs in freedom away across the field.

Well this difference between the narrator and Mahony, then, turns out to be a relatively subtle one and is connected to the idea of thwarted or frustrated expectations which is acted out in the plot.

We might recognize, then, that one of the most fundamental things that's going on in the story is that Joyce (we might move from our description to a slightly more generalized level, but it's grounded in our description of the actual events of the story), what we recognize is that the story is partly about quavering intellection. It does something that Joyce does again and again, and it's a rather surprising thing for a writer as intelligent and as learned as James Joyce himself seems to be, because what he's dramatizing is the dependence of all of us on ordinary humanity.

The great epiphany at the very end of the story has the narrator calling to Mahony across the field, and the last lines of the story run something like this: "He ran as if to bring me aid," the narrator writes, "and I was penitent, for in my heart I had always despised him a little." In other words, the narrator has learned a lesson. The lesson has been how he's dependent on this ordinary guy, that brains are not the only qualities in a human being, that high intelligence does not protect one from being cowardly sometimes, or being foolish, or sometimes opening one to corruption that other more active, spontaneous, and younger people might be oblivious of or freed from.

So the story ends dramatizing the narrator's recognition of his own dependence on and respect for a boy that he had previously despised. The limits of intellection is one of Joyce's subjects. This is also one of the deep, deep subjects of Joyce's *Ulysses*, in which the central character, the hero of the book, the protagonist of the book, is the most ordinary of men, and in which the character who's most like Joyce—Stephen Daedalus, the character who comes into *Ulysses* from the autobiographical novel, *A Portrait of the Artist as a Young Man*, comes across in *Ulysses* as something of a prig, as something of a self-centered narcissist, and as certainly lacking the wideness and generosity that Leopold Bloom, the ordinary protagonist of the novel, shows.

Well, there's a complex symbolism in "An Encounter" as well, but even the symbolism is grounded in the manifest obvious central parts of the story, and I sometimes in my classes will pause and talk briefly about the symbolism. And the students often feel that symbols are arcane, difficult things. How can they figure them out? And the answer is the same as the answer I've been giving to earlier questions. What is symbolic in a story will be signaled by the story.

If you stop and think for a minute you'll realize that because fiction is so full of items of furniture, of items of clothing, of debris, that if you were under the impression or under the responsibility to find symbolism every place in the story, the stories would become unreadable, because potentially every single object in the story might have a kind of larger symbolic significance. Of course, the answer is that's not true.

Items that have a significance, that have a larger significance than is normally attributed to them—which is what a symbol is: a symbol is an item in the story that is granted or to which is attributed a larger meaning than would normally be the case—those items in the story that have that valence, that have that symbolic charge, are signaled to us by the story.

And in the case of Joyce's story the symbolism has to do with green eyes. At an early point in the story the boys walk down to the docks, and the narrator says, "I was looking in the eyes of the sailors to see if any of them had green eyes because I had some confused notion...." The green eyes are a symbol for the boy of the exotic, of the adventurous, of freedom, of escape. Where does he encounter the green eyes? In his encounter with the pervert. "I looked up into the face and saw a pair of bottle-green eyes." So his symbol of the exotic is discovered in the face of corruption.

What have we discovered here? That the symbolism in the story reinforces what we've already learned about character, and reinforces what we've already learned about plot—and also that the symbolism in the story is obvious. Anyone reading the story would recognize that the green eyes have a larger significance than they might have in another kind of story where they simply signified a character's eye color.

I need to say some final words about the style of the story, about the language in which the story is written. Remember how I said earlier that if we describe a story's constituent elements carefully, we can get to the meaning of the story; and I talked about plot, character, style. We said a few words about symbolism as well, to complicate our argument. But let's say a few words about style, because the contribution of Joyce's particular language is crucial to our understanding of the story's subtlety.

One way to understand what Joyce does in the story is to recognize that he establishes a particularly delicate balance of sympathy and judgment for his narrator. The sympathy is created by the first person narration. We get a view of the character in some sense from the inside, and it establishes an intimacy between us and the character that no other point of view could possibly do. But in addition to that, Joyce complicates that by his vocabulary; because it's the vocabulary not of a child, but of an older man looking back on his earlier self with a kind of amused, detached irony.

Listen to this sentence: "When the restraining influence of school was at a distance, I began to hunger again for wild sensations, for the escape which those chronicles of disorder seemed alone to offer me." Well, to describe a 10-year-old boy's dream of adventure as a hungering for wild sensations, and to call his favorite books "chronicles of disorder," that is to expose the boy to a gentle, amused mockery. It's a kind of affectionate mockery. It's tolerant; it's amused; but nonetheless it judges the boy's innocent foolishness.

There are many, many other instances of this complexity, of this quiet complexity of tone, in Joyce's diction, and what this reminds us of is that the story concerns but a minor incident. Imagine what might happen if the boy told his own story the day after the event. "Oh, my world is over. I thought the world was beautiful, and I realize now it's full of perverts."

But because the narrator is telling the story 20 years later, or 30 years later, looking back on his youthful follies with a kind of amused irony, we are reminded of how much more complex the world is, that it takes many lessons to make an adult. And we are also reminded by the narrator's austere vocabulary and syntax, by his whole gently ironic tone, not only that it takes many lessons to make an adult—that what we are reading is not *Paradise Lost*, but a single short story—but also that the narrator's good sense and tact and humanity are essential qualities also for readers of literature and far more valuable than a gift for puzzle-solving or a specialist's trained eye.

Lecture Three
Defining Modernism—Monet's Cathedral

Scope:

The term *Modernism* has been problematic since its coinage by the generation of artists and writers who first used it to distinguish themselves from their 19th-century predecessors. For the purposes of this course, Modernism designates the experimental and avant-garde movements in the arts and literature that emerged in the early 20th century in Europe and America from roots in French painting and elsewhere. We will briefly highlight some of the defining philosophic, scientific, and cultural features of the era to establish a context for the stories and novels that emerged during this time. We end by examining some classic paintings by Monet.

Outline

I. The term *Modernism* has multiple meanings depending on who is defining it and when it is being defined.
   A. Complicated by academics and others, Modernism can mean different things in different fields.
   B. Every age thinks of itself as modern. Worse, the term *Postmodern* is even more unhelpful, despite its widespread and confusing deployment in contemporary culture.
   C. Modernism in the context of this course addresses the experimental and avant-garde movements in arts and literature that emerged in late-19th- and early-20th-century Europe and America.
      1. The rise of Modernism is rooted in changes in French painting and other disciplines that arose during the period of Romanticism at the turn of the 19th century.
      2. For our purposes, Modernism designates the period roughly between the 1880s and 1930s. Modernism as a conscious artistic and literary project appears earlier in Europe than in England or America.
   D. No major shift in the arts is generated from within; the arts reflect politics, philosophy, economics, and world events.

E. Modernist writers and painters were part of a much larger cultural and intellectual transformation whose essential nature is implied, in part, in the unsettling skepticism, nostalgia, and moral ambiguity we glimpsed in the poem "Captain Carpenter" in Lecture One.

II. A summary of key developments in economics, philosophy, and history will provide a cursory background to Modern literature.
   A. For Karl Marx, economic forces and class struggles shaped consciousness.
      1. The idea that humans can be victims of forces beyond their control formed the heart of his work.
      2. This attitude was also widespread among other reformist writers such as Charles Dickens, H. G. Wells, and George Bernard Shaw.
   B. Skepticism toward inherited categories and notions of philosophic coherence became a central feature of the later 19th century.
      1. For example, the influential philosopher Arthur Schopenhauer was known for his skepticism and pessimism. He believed that forces beyond or within us operated coercively to shape individual consciousness.
      2. The philosopher Nietzsche is associated especially with an iconoclastic, pessimistic, even intellectually revolutionary attitude toward inherited notions of value and coherence.
   C. Challenges to the received wisdom in philosophy and moral discourse were replicated in science, where, for example, Darwin's theory of evolution displaced man from the center of the universe.
   D. Einstein's special theory of relativity (1905) showed that time and motion are not absolute but relative to the observer, thus powerfully upsetting physical laws thought to be unchanging.
   E. In his *Ethics* (1912), G. E. Moore tried to respond to the emerging moral relativism of the turn of the 20th century. If what we are coming to know about the world is true, where does the notion of ethical values come from?

- **F.** Ludwig Wittgenstein and Bertrand Russell pioneered analytic philosophy and developed a new skepticism toward language, destabilizing what had been thought of as a fixed relationship between words and meaning. Some of the narrators of Modern fiction embody a profound version of this skeptical idea of language and the difficulties of communication.
- **G.** Sigmund Freud also destabilized inherited notions of the self with his idea that "we are driven by forces" we do not understand or even acknowledge to exist: the unconscious, issues of sexuality, and trivial events in everyday life.
- **H.** William James, brother of novelist Henry James, further complicated our ideas of human nature (and coined the term *stream of consciousness*).
    1. He believed that states of personal consciousness are in constant flux.
    2. This theme is picked up by many writers, especially D. H. Lawrence and Virginia Woolf.
- **I.** Add to this moral, psychological, and scientific transformation, the physical transformation of human life in the advanced industrial societies that occurred in the 19th century: the invention of machinery and new technologies in work, transportation, and communication.

III. Some of the fiction we will study dramatizes the sense of disconnection that arose from the upheaval of the late 1800s. But there is a less despairing side to Modernism, as we can see by considering the great Impressionist and Postimpressionist painters whose works prefigure Modern fiction.
- **A.** Let's close by considering Claude Monet, one of the great founding Impressionists. His signature works are series-paintings of the same object in different light, time of day, climate—for example, poplars and haystacks, the Paris train depot, images of London, water lilies, the gothic cathedral in Rouen, France.
    1. Monet's 32 images of the cathedral show how its appearance changes constantly, resisting capture. The series embodies the idea that the cathedral must be represented in its multiplicity and endless nuance.

2. The immense complexity and uniqueness of each instant is part of the paintings' meaning, as is our own subjectivity; we half-create the things we see.
   3. The visible world is still beautiful in this Impressionism, but it is evanescent, transitory, and especially, unique to each perceiving eye.
B. Even in the Rouen Cathedral series, however, we can see a strong tendency away from simple representation and toward a more purely "painterly" interest in form and color for their own sake.
C. Monet's awareness of the painting as a painting constitutes a self-consciousness that came to be understood as a defining feature of the Modern. This idea has crucial echoes in the novelists we'll discuss.
   1. One way in which the Modernists differentiated themselves from their predecessors is that they rejected a strictly realistic representation of the world.
   2. They, like Monet, found ways to register their sense of art's limitations.
D. Monet's water lily paintings, done in his last 20 years, clearly dramatize a drive toward the abstract, toward a "pure" painting almost—but not completely—devoid of representational elements.
E. This tendency to move away from realistic representation toward symbolic or abstract forms is a powerful force in Postimpressionism; and this principle is reflected in more problematic and disturbing ways in some of the fiction this course examines.

Essential Reading:

Hamilton, *Painting and Sculpture in Europe, 1880–1940*, pp. 34–41.

Supplementary Reading:

Ellmann and Feidelson, eds., *The Modern Tradition: Backgrounds of Modern Literature*, excerpts on Schopenhauer, pp. 545–547; Marx, pp. 741–744; Freud, pp. 559–571; William James, pp. 715–723.

Riding, "Monet's Fixation on the Rouen Cathedral," *New York Times*, 15 August 1994, pp. C9–10.

Questions to Consider:

1. What assumptions about reality are implied by Monet's practice of painting the same subject over and over again?
2. How might this practice be related to emerging scientific and psychological notions of instability and subjectivity?

Lecture Three—Transcript
Defining Modernism—Monet's Cathedral

The term "Modernism" has been problematic since its coinage by the generation of artists and writers who first used it to distinguish themselves from their 19th-century ancestors. The term has been immensely complicated since by academics and even by public-policy specialists. The architectural historian's "Modernism" is very different from the economic and political "Modernism" referred to in our current discourse on globalization.

So it's a bad term. It can mean too many things. It's very widely used. And it's also a bad term because every age thinks it's "Modern." I guess the term "Postmodern" is even more unhelpful from many angles, despite its widespread and confusing deployment in contemporary cultural argument.

For the purposes of this course, let's let "Modernism" designate the experimental and avant-garde movements in the arts and literature especially that emerged in the late 19th and earlier 20th centuries in Europe and America, from roots in French painting and elsewhere. In Romanticism, in the Romantic Movement of the earlier part of the 19th century, the turn of the 19th century, also a crucial root, a crucial source or origin for Modernism.

Diverse in the extreme, this body of material nevertheless shares certain features which our course will explore in detail. For us, then, the term is first a period designation, marking out a broad period, roughly the period between, let's say, the 1880s and the 1930s. Modernism as a conscious artistic and literary project appears earlier in Europe than in England or the United States, and that's one reason our dating must remain vague and encompassing.

But no major shift in the practice of the arts is generated from within. The arts are not an autonomous environment, an autonomous energy within culture. They reflect cultural, and philosophic, and political, and economic realities. They as much reflect what is going on around them as they do help shape those currents, those pressures. And the often radical, deeply experimental painting and literature we call Modernist is no exception to this principle.

The Modernist writers and painters were part of a much larger cultural and intellectual transformation, whose essential nature is implied in the unsettling skepticism, balked nostalgia, and moral ambiguity that we've already glimpsed in the poem "Captain Carpenter."

What I'd like to do in this lecture is provide a kind of context within which to understand Modern literature, a kind of very skeletal background in which we take account of other forces that lie behind the Modernist writers and help to explain the particular character of Modernist literature.

Now, many of the materials I'm going to talk very briefly about in this talk could themselves be the center of their own subject, of their own course, and in fact many of them are, in university and elsewhere. The courses in history, in economics, in philosophy, or in cultural history might center on the figures I'm about to talk about and leave the writers that we will focus on in this course more in the background. What I'm offering here then is a simplified and skeletal map of what we could call the surround of Modern literature, the background to Modern literature.

We might begin by mentioning a figure whose importance to the history of culture has been a source of disagreement and controversy almost from the beginning: Karl Marx. For Marx, history is a kind of engine that obliterates the old pieties about human agency. Economic forces, class struggles shape consciousness for Marx. The messianic Marx, the Marx whose thought was seized upon by totalitarian Communism, may belong to religion and to totalitarian Communism, but the man who mapped the birth of capital changed our way of understanding experience.

The idea that human beings are in some sense, in good part, the victims of forces beyond them, forces too large for them to control, of course is at the very heart of Marxist thought. And we can get some sense of the way in which this kind of attitude was not simply restricted to Marx but became very much more widespread.

If we compare reformist writers like Dickens in the earlier part of the 19th century to writers in a later era like 20th century writers like Shaw or H.G. Wells, these are all sort of social realists who are concerned with the role of individuals in connection or against the backdrop of the larger social order. And in the case of these writers what we feel is that in Dickens and in other writers like him there's a much greater emphasis on the notion of individuals changing themselves, transforming their lives by acts of will and by acts of individual reformation. They reform themselves.

But the later writers are much more conscious of massive social and historical energies that control human life, or that shape it and contain it, and it's those larger forces that the later writers are aware of as requiring change and requiring reform.

There's a philosophic background to these attitudes as well. Some of the major philosophers and most influential philosophers of the 19th century wrote philosophic arguments that essentially undermined or put in question the inherited categories and notions of philosophic coherence that had come down to their time.

One of the most important of these is the philosopher Schopenhauer, whose philosophy is often described as a kind of Modern skepticism. One of his titles is *The World as Will and Idea*, and the notion that the world itself, the external forces of the world, operate as a kind of will, a kind of coercive force that shapes individual consciousness, is at the heart of his pessimism, is at the heart of Schopenhauer's view of experience.

The philosopher Nietzsche is especially associated with a kind of iconoclastic, pessimistic, aggressively, even intellectually revolutionary attitude toward inherited notions of value and coherence.

And these philosophic and historical conceptions of a world of experience that presses upon individuals and controls them in certain ways had a kind of counterpart in developments in the birth of modern science.

If we think of the implications, for example, of Darwin's theory of evolution, which emerges centrally in the 19th century, we can see how what the theory of evolution does is displace man from the center of the universe; and, of course, many religious people were very disturbed by Darwinian theory exactly for this reason, and continue to be. An interesting thing that even into the 20th century, and now into the 21st century, arguments over Darwin's theories center on questions having to do with whether or not there is a prime mover or a God. So, Darwinian theory also displaced man from the center of the universe and laid emphasis on large, long-standing forces, shaping forces.

Einstein's work begins to appear in the early part of the 20th century. His famous paper, "The Electrodynamics of Moving Bodies," was published in 1905 and later came to be known as the special theory of relativity. And in this theory, both time and motion are understood not to be absolute, but relative to the observer, as if even our most fundamental conceptions of time and physical motion are put in question, are rendered problematic by the developments of modern science, by the emerging notions of science.

And it's as if the physical universe itself had been relativized by such arguments, and that notion of relativism begins to have a counterpart in the world of ethics. Moral relativism becomes a crucial topic for argument and discourse at the end of the 19th and in the early parts of the 20th century. And the English philosopher G.E. Moore wrote a book entitled *Ethics*, published in 1912, trying to respond to and answer the moral relativists who were saying, "Look, if what we're coming to know about the world is true, where do ethical values come from?" Again, a deeply destabilizing principle that makes inherited moral categories much more problematic than they'd been.

The great language philosophers Wittgenstein and Bertrand Russell begin to develop theories of language, which make language itself more problematic than it had been before. Wittgenstein, the great philosopher of linguistic complexity and limitation, wrote at one point in his famous *Tractatus* (1921): "Language disguises thought. So much so, that from the outward form of the clothing it is impossible to infer the form of the thought beneath it,"—an idea that language, far from being an instrument for understanding, might be an instrument for obscuring our understanding of the world.

Some of the driven, even compulsive narrators of Modern fiction might be said to embody a profound version of this skeptical idea of language and of the difficulties of communication. And Sigmund Freud's work is part of this process of destabilizing and problematizing the sense of the world that was inherited by and was in some sense shared by the Modern novelists we'll be studying, because Freud's is another skeptical and destabilizing vision.

In Freud's conception of the world "we live by forces that we don't fully understand" or even acknowledge, even know exist, in fact. The theory of the unconscious, the importance of sexuality, the significance of trivial or minor things in our everyday experience, all these Freudian notions complicate and problematize inherent attitudes about personality and about human selfhood.

There's an American contributor to this project who's worth mentioning, complicating our ideas of human nature very deeply. I'm thinking of William James, brother of the great novelist Henry James. William James coined the term "stream of consciousness." It's become such a central category for Modern fiction. It's the title of one of the chapters in his book *Principles of Psychology* published in 1892. And he says in that book, "Within each personal consciousness states are always changing … each personal consciousness is sensibly continuous."

Well, this idea that the self itself is unstable, that identity is not fixed but something in motion, something becoming, is in many ways a disturbing notion. It certainly complicates and destabilizes inherited conceptions of the relation between self and world and the relation between selves and selves, and it is a notion that is picked up and deepened and complicated in many of the Modern writers we'll be looking at, most especially in writers like D.H. Lawrence, and, as we will see, Virginia Woolf.

Well, I've been describing some of the sources for the debstabilizing skepticism and self-consciousness that become signature features of Modern literature. History, philosophy, notions of ethical value, language, and even personality or the human mind itself are understood to be newly ambiguous, fragmented, disorderly.

And let's add one other even more obvious fact to this—the physical transformation of human life in advanced industrial societies that occurred in the 19th century. The way ordinary people encountered and negotiated with the coming of the machine, the assembly line, the new technologies of transport and communication. Some of these will define 20th century experience. A new world of speed and steel. A mobile world where people leave their homes and their childhood places for cities, other countries, other continents. People are less connected with their ancestors and with the communal traditions that had sustained them than any human beings before them.

So, this sort of basic view of destabilized circumstances into which our Modernist fiction must fit, I think is important and helpful. But it's a mistake to see these developments as entirely bleak, as leading inevitably to the despairing nostalgia that we might associate with the poet T.S. Eliot. Some of the fiction we'll be studying will show this kind of darkness very dramatically, but we can glimpse a less despairing side of the Modern, also to be reflected in our novelists, most easily I think if we think of the great Impressionist and Postimpressionist painters whose project may be said to parallel that of Modern fiction.

So, I want to conclude this lecture by turning to that quintessential Impressionist, Claude Monet. He was born in 1840; died in 1926, a figure deeply of the 19th but also of the 20th century. All his work, the full trajectory of his long career, is relevant, of course, but we can come close to his essence, and to the essence of his importance for our understanding of literary Modernism, by recalling his famous series-paintings.

There's a famous series of poplars and haystacks, another of the Paris train depot. Another series, very lovely series that the English especially treasure, of London and of the Houses of Parliament, and of the River Thames especially in fog. Monet loved fog and what fog did to the physical presences of buildings and trees and so forth; and, of course, the great series of paintings of water lilies, to which Monet devoted the last 20 years of his life.

Why would Monet paint the same items again and again? What's implied in his preoccupation with the way trees or haystacks or water lilies alter depending on light, on the time of day, on the season, even on the painter's own vantage point or perspective, perhaps even on his emotional state?

Well, I think we can seek the answers by focusing on what is for me, Monet's greatest series: the 30 astounding paintings of Rouen Cathedral that he made at two different times—in February and in April 1892, and then again some months later in 1893. Monet produced 28 images of the main façade of the 14th-century Gothic masterpiece, the Cathedral of Rouen. I said he made 30 paintings, and he did.

The only two oil paintings not of the main façade show medieval houses that had been attached to one of the cathedral towers. These medieval buildings were destroyed by the Allied bombings of World War II, and it made those two paintings especially precious. Those two paintings as well as 14 others of the main façade were brought back to Rouen in 1994 for an exhibition to commemorate the hundredth anniversary of the series of paintings. The paintings are officially dated 1894, because after he began them and started them, Monet took them back to his home and finished them at home.

"My cliff, my cliff," Monet complained, as he worked simultaneously on as many as 14 canvases at the same time, making minute changes as the light altered, aiming to capture the instant before it disappeared. One amazing aspect of the series is how changeable the cathedral seems—solid and clear in one image, appearing to float and recede in others; suffused in fiery colors in some canvases, and then in fog or cloud in others. Slight shifts in the painter's angle of vision create stunning variations or changes in the play of light and color across the cathedral's sometimes spectral, upthrusting western façade. The constantly changing light and unpredictable weather of Normandy is one of the implicit subjects of the series.

They are absolutely amazing paintings. Reproductions don't do justice to the quality of surprise you feel when you look at one painting and then at another next to it, and in fact I've come to realize that it's one of the tragedies of modern life that the paintings have never been shown together. Almost from the first moment that they were exhibited they were sold off in different ways, and there are two magnificent versions of the paintings in the National Gallery in Washington, D.C. There are wonderful versions of the paintings in museums in New York, and of course others of the series are in various museums in Europe. But even that exhibition in Rouen only brought together 16 of the paintings, and it would be an astonishing act of cultural intelligence and respect for the history of art if some museum brought all 30 together. The experience of seeing them all together I am certain would be a revelation to many viewers, even to many people who were very familiar with the paintings.

Reality, these paintings say, is not one fixed thing; it constantly changes. It is unstable; it resists capture; it must be represented in its multiplicity and its endless nuance. The immense complexity, the uniqueness of each instant, is part of the meaning of these great, tentative paintings. So is our own subjectivity a part of the meaning of the paintings. We half-create what we see. "We half-create the things we see," Monet's paintings seem to say. The visible world is still beautiful in this Impressionism, but it is evanescent, transitory. In part, it is even unique to each perceiving eye.

There are profound counterparts to this insight in many of the literary texts we'll be studying in this series of lectures. Perhaps the most dramatic and immediate connection between Monet's idea and the Impressionist and Postimpressionist idea of reality is in Virginia Woolf's novels, and especially in *To the Lighthouse*, which we'll be talking about down the line later in the course.

There is a central character in *To the Lighthouse*, a woman named Lily Briscoe, a painter, a counterpart to Virginia Woolf herself, who through the whole of the novel is working on a painting. The novel takes place over a long period of time. The first part of the novel takes place in an afternoon. The third part of the novel picks up 10 years later, and Lily picks up her paintbrush and begins to work on and complete the painting that she'd been working on earlier.

In her descriptions of Lily's ambition for the painting, Woolf also articulates her own ambitions for her novels. And the descriptions of Lily's ambitions for her paintings are drawn, one almost feels that Virginia Woolf was looking at paintings like those by Monet or Cézanne in order to conceive the aesthetic arguments she was making about the evanescence of the moment.

The visible world as I've suggested is still beautiful in the Impressionism we see in Monet, but it's evanescent, transitory, unique for each perceiving eye. Monet's career itself can be said to enact something of the larger story of modern art beyond Impressionism. He outlives all his contemporaries and most of the generation of Postimpressionists who are said to build on him and also to reject his psychological realism. Prime instances of this later generation would be Cézanne, van Gogh, Picasso; and I'll say a few things about these painters in our next lecture.

But even in the *Rouen Cathedral* series one can see a strong tendency away from simple representation, toward a more purely "painterly" interest in form and color, in the materials of paint for their own sake. The swirl of fog or sunlight against stone comes to own the paintings sometimes in the *Rouen Cathedral* series. The cathedral becomes an instrument of verticals and slender jagged-edged cylinders or cones, played across by light and shade and color.

This awareness, in Monet and other Impressionists, of the painting *as a painting*, of its physical existence as canvas and oil paint, constitutes a profound kind of self-consciousness that again has crucial counterparts, has crucial echoes in the novelists we'll be talking about. This kind of self-consciousness will come to be understood as a defining feature of "the Modern."

One way in which the Modernists saw themselves as profoundly different from their ancestors is connected exactly to this idea that they have a kind of what they would call honesty about the project they're engaged in. What you're looking at is not a photographically realistic representation of the world. You're not supposed to forget the painterly strokes or the frame surrounding the painting when you look. It's not as if the painting is a window on the world and the medium is a kind of transparent medium from which reality simply emerges, or in which reality is simply reproduced. It's something much more complicated than that. The medium that is looking at the reality changes the reality, affects the reality, is not the same as the reality.

So, when we talk about the self-consciousness or the self-reflexiveness of these paintings or of Modern fiction, what we're talking about in part is the artist's attempt to remind the viewer, to remind the reader that what you're looking at is words on a page, that what you're looking at is paint, that reality itself is something beyond, that you're looking at a representation of the world, not at the world itself, and that this difference, this mediation between world and representation of the world has to be a part of your understanding of the experience of art. It's a form of honesty, a form of candor, even, in a certain way, a confession of limitation.

So this awareness in the painters, especially in Monet and the other Impressionists, of the painting as a painting is a crucial element and has powerful counterparts in our Modern fiction.

The water lily paintings that I mentioned before, painted and posed in Monet's backyard pond—his canvas really, for 20 years—clearly dramatize this drive toward the abstract, toward a "pure" painting, almost a pure painting—*but not quite!* A pure painting almost devoid of representational elements, but not completely. You can still recognize something of the object represented.

But this tendency to move away from realistic representation toward symbolic or abstract forms in which the realistic or representational element becomes only vestigial is, of course, a powerful energy in Postimpressionism and in the abstract paintings that grow out of

Postimpressionism. And we can see this principle also reflected in some of our fiction, in more problematic and disturbing ways in some of our fiction.

In James Joyce's *Ulysses*, for example, as the narrative level of the story kind of recedes as the novel goes on, and as the various voices and perspectives that Joyce introduces to describe the world of his book proliferate, you begin to feel that there's a kind of performance going on in which the medium of language becomes the subject of the book more than the story of Leopold Bloom and his wife. Not quite. You are still interested in Leopold Bloom, and the realistic narrative retains its importance; but it recedes in some sense. It no longer occupies the central place that it did earlier in the novel. And this progress toward a kind of increasing abstraction and self-consciousness about the medium itself is characteristic of Modernist literature and Modernist art.

And this is especially true of Monet's *Water Lilies*. In some of the late water lily paintings the viewer cannot distinguish the flowers from the fluid element in which they swim or float. The border between solid and watery dissolves. The painting becomes an expression of color, form, mass, volume, almost an abstract meditation on the nature of paint, on the conditions of paint.

But at this point, of course, biographical and mortal truths intervene, for we need to remember that these last immense blobbish wonders are astonishing paintings. Some of them can be seen today in New York City at the Museum of Modern Art. But these paintings were painted by a half-blind, 86-year-old, frail, doing those lilies once again. He'd been painting them for 20 years; doing those lilies yet again, the brush taped or tied with rags to his stiffened fingers; a great hero of art.

The immense, complex grandeur of what we take for granted, often that we never even notice, is one meaning of Monet's work, and especially one meaning of the *Rouen Cathedral* paintings. Describing both the simplicity and the grandeur of Monet's project in the *Rouen Cathedral* series, one scholar put it memorably: "In 1892 Monet focused on the evening light. In 1893 he went after the morning."

Lecture Four
Defining Modernism—Beyond Impressionism

Scope:

"[O]n or about December 1910 human character changed." So said Virginia Woolf in her 1924 essay, "Mr. Bennett and Mrs. Brown." This lecture explores part of what Woolf meant when she described Modernism as this radical break with the past. Our discussion continues to examine visual art as a context for the literary material we are preparing to read. A brief sketch of Postimpressionist art aims to describe a difference between early and later Modernism—between, say, Conrad and Joyce or Kafka. In this later art we see a greater degree of distortion, a movement away from representational realism toward forms of expression that try to reveal inner energies in nature and in the observer-painter. In Van Gogh's *The Starry Night*, we can feel the artist's desperation as he tries to paint this vibrating nightscape as it changes before his eyes. That the world escapes the artist even as he creates the painting is a profound aspect of the experience of looking at this remarkable work of art. We will see that same agony and ecstasy in some of the Modernist writers studied in this course. We end by comparing the voices of fictional narrators from the 19th century with their 20th-century counterparts—a movement from confidence and omniscience to tentativeness, self-consciousness, inwardness, uncertainty.

Outline

I. Woolf's essay alludes to larger matters but refers also to a specific event: an exhibit of Postimpressionist paintings organized by her friend Roger Fry in London (1910–1911).

    A. Although generally reviled, the show triumphed over its mocking reception and opened England to the tides of Modernism.

    B. One writer expressed a widespread revulsion: these were "…works of idleness and impotent stupidity, a pornographic show."

1. It may seem surprising today that an art so stylized and tame should have aroused such hostility and offense. But the representation of human bodies was a radical act.
2. One of the deep projects of Modern literature and art has been to examine our lives in our bodies with a vigorous, unsentimental candor that was unavailable to previous generations.

C. The show included Cezanne, Van Gogh, and Gauguin. Different as these and other Postimpressionist painters are, they share at least one defining idea: that exaggeration or distortion may lead us to deeper or more compelling truths than the exact representation of physical or outward reality.
1. In looking at Monet's paintings in Lecture Three, we saw that this principle was deeply embedded in the first generation of French Impressionists.
2. But in the next generation, realistic representation remains an element but declines in importance and competes with the painting's "self-reflexive" interest in form, color, and volume.

II. The distorting element in Impressionist art grows more obtrusive and central in Postimpressionist works. The painter Matisse formulated an influential distillation of this principle: it is necessary to sacrifice exactitude to truth. (The French, "L'exactitude n'est pas la verité," is more precisely translated as "Exactitude is not truth.")

A. One path for this principle led to Surrealism; another led to Cubism and abstract art.

B. Marcel Duchamp's *Nude Descending a Staircase No. 2* is one famous embodiment of Cubist principles—a naked woman represented as a series of geometric shapes, machined and metallic.
1. The Modern is also the era of machines, of transforming new technologies and mechanizing forces.
2. The painting collapses or manipulates time, representing motion and duration simultaneously. This manipulation of time and the impulse toward abstraction will be signal features of many of the literary texts in this course.

- **C.** We see another example of these principles of distortion or heightened representation in Picasso's portraits of Dora Maar, in which the artist's mistress becomes a distorted, mechanized geometric figure.
- **D.** Distortion is also a key aspect of the satiric, disgusted images of depravity and nightmare created by the German Expressionist painter Max Beckmann. His painting, *Family Picture*, shows a grotesque gathering of people in a room.
    1. The perspective of the work is altered, and the figures and objects seem cramped, pressed together, and constrained.
    2. None of the figures seems aware of anyone else.
    3. There is an atmosphere of nightmare and sickness.

III. But not all Postimpressionist work is dark and despairing.
- **A.** Van Gogh's *The Starry Night* is a less bleak and perhaps more representative Modernist work.
- **B.** The painting displays an immense self-awareness.
    1. We feel the painter's desperation to capture a reality that disappears in the instant of its occurring: the uniqueness, the fluidity—the evanescence but also the incredible beauty—of the moment.
    2. That the world is escaping Van Gogh even as he is creating the painting is a profound aspect of our experience of this remarkable work of art.
    3. The agony and ecstasy of this astonishing picture have a counterpart in text we will read by Joseph Conrad, Ford Madox Ford, James Joyce, D. H. Lawrence, and Virginia Woolf.

IV. One answer to the question of the larger meaning of Woolf's remark that human character had changed can be found by comparing 19th- to 20th-century fiction, and in particular by comparing the voices we hear in the opening paragraphs of some famous novels from each century.
- **A.** Listen to the contrasts, the decisive differences, in these comparisons:
    - Jane Austen's *Emma* (1815) and Joseph Conrad's *Lord Jim* (1900)

- William Makepeace Thackeray's *Vanity Fair* (1847–1848) and William Faulkner's *The Sound and Fury* (1929)
- George Eliot's *Adam Bede* (1859) and Joseph Conrad's *Under Western Eyes* (1911).

B. Each of the earlier works has a confident, all-knowing narrator. The 20th-century narrators, in contrast, speak hesitantly and insist on the limits of what they've seen and understand. They give the reader much less help, no exposition, and demand a more strenuous attention than older storytellers.

Essential Reading:

Hamilton, *Painting and Sculpture in Europe, 1880–1940*, pp. 41–49; 94–103; 235–253; 474–484.

Woolf, "Mr. Bennett and Mrs. Brown," in *Approaches to the Novel*, Scholes, ed., pp. 211–320.

Supplementary Reading:

Hynes, "Human Character Changes," in *The Edwardian Turn of Mind*, pp. 307–345.

Frank, "Spatial Form in Modern Literature," in *The Widening Gyre: Crisis and Mastery in Modern Literature*, pp. 3–62.

Questions to Consider:

1. What does Matisse mean by his paradoxical assertion that it is necessary to sacrifice exactitude to truth?
2. What do Modern novelists gain by surrendering the omniscient perspective of their predecessors in the 18th and 19th centuries? What do they lose?

Lecture Four—Transcript
Defining Modernism—Beyond Impressionism

In an essay published in 1924 titled "Mr. Bennett and Mrs. Brown," Virginia Woolf wrote the following sentence; it has become one of the most famous sentences in the critical history of Modernism: "On or about December 1910 human character changed." What did she mean, and why this specific date?

Let's leave the big first question implicit for the moment, and answer the second. Why December 1910? The quick answer, the most fundamental answer, is that from November 8, 1910 through January 15, 1911, there ran in London an exhibition of Postimpressionist paintings. It was reviled at its opening by an immense range of respectable artists and intellectuals as well as the general public, which was disgusted by many of the images that were shown there. But the show triumphed over its mocking reception, opening England to the tides of Modernism, and securing the show's place in the history of the Modernist movement.

One example of the intense hostility that these then-experimental paintings generated can be shown from this passage in a diary written by a literary and artistic intellectual of the time, Wilfrid Blunt. He spoke in his diary entry of that "gross puerility which scrawls indecencies on the walls of a privy," and compared that to the paintings. He said, "These are not works of art at all. ... They are the works of idleness, impotent stupidity, a pornographic show."

It seems amazing to us today from this vantage point in the 21st century that an art which seems in some way so stylized and certainly tame in comparison to the pornography we can click on on the Internet or even bump into by going to the local newsstand, it seems astonishing today that these materials would have aroused such hostility and especially such offense at the representation of human bodies. But as I'll try to indicate a bit later, the representation of sexuality, and the representation of our lives and our physical bodies, is one of the central projects of Modernism, and is one of the central reasons to value Modernist literature especially.

It's often forgotten that Modern literature and art was deeply committed to examining our lives and our bodies with a vigorous, unsentimental candor that previous art did not manage. You might think, in trying to imagine what this Postimpressionist exhibition in

London was like, you might think of some of the famous paintings of naked bathers, although they are very stylized paintings if you've seen them, by various Postimpressionists, especially Picasso and Cézanne, both of whom did very famous paintings of bathers, of women bathing.

The show I'm describing—the Postimpressionist exhibition—was organized by Virginia Woolf's friend, the art critic Roger Fry, and the show itself is actually described in some detail in a biography that Woolf herself wrote about Fry, published in 1940. The show was titled *Manet and the Post-Impressionists*, and it included paintings by Cézanne, van Gogh, Gauguin, and some of the younger Postimpressionists as well, including Matisse and Picasso.

Different as these and other Postimpressionist painters are, they share certain features, and in particular, one great defining principle, which will have a decisive counterpart in the literary texts we'll be studying in these lectures. This is the principle that exaggeration or distortion may lead one to a deeper or more compelling truth than the exact representation of physical or outward reality.

Now, of course, we've already seen how this is a principle deeply embedded in the first generation of French Impressionist paintings, and I was trying to indicate something of that when I spoke about the way in which the trajectory of Monet's career leads toward a greater and greater degree of abstraction, in which the realistic representation remains present in the painting but declines in importance and competes with the painting's interest in form and color and volume.

But perhaps one could say that this distorting element grows more obtrusive and more central in the Postimpressionists. One sort of motto or distillation of this principle comes from Matisse, inspired by Cézanne who said: "Exactitude is not truth," an apparently paradoxical statement, but the truth of which we will see reflected again and again in some of the literary texts we'll be looking at.

One strand or path for this principle of exactitude being sacrificed to truth leads to the painterly movement called Surrealism, and there are a number of possible choices one might make to embody these principles. But maybe a touchstone or a representative piece—in part because it reflects not only the Surrealist Movement, but even more the influence of a movement that came to be called Cubism, and even the related movement that was linked to Cubism and grew out of it in and soon to be called Futurism—the touchstone for our purposes to illustrate this principle might be the famous painting by Duchamp called *Nude Descending a Staircase*, exhibited in 1912. Almost every 20th- and 21st-century human being has an optical memory, I think, of this famous and in many ways strange painting. What's of special note in the painting are a couple of things, I think. The recognizable nude descending the staircase virtually disappears, and it seems as if what Duchamp is trying to represent are the successive movements of a single body. It's likely that he was influenced by the multiple exposure photography documented by Eadweard Muybridge in the precursor of the movies in the famous book *The Horse in Motion*, and it's likely that the painting was influenced by that book in certain ways, and by Muybridge's example.

We note if we look at the painting an emphasis on the geometric, the machined, and the metallic. It's important to remember that the Modern period is the era of machines and of transforming new technologies and mechanizing forces, and that's partly reflected in the *Nude Descending a Staircase*. Maybe most important, however, is this attempt to represent simultaneity, the collapsing or manipulation of time in the painting. The nude is in motion not only in space but also in time. The artist is attempting to represent or gesture toward the representation of motion and duration simultaneously. The painting wants to show us something that aims for a kind of simultaneity. And this is linked, of course, to its profound tendency to abstraction.

These impulses, especially the impulse to manipulate time and this abstracting impulse, will be a signal feature of many of our literary texts as well.

We might also, to illustrate this principle, look at a range of Picasso paintings. Almost any of Picasso's middle and late paintings would fit the bill here. The Dora Maar portraits are a particularly good example, in which we see Picasso famously representing his mistress in ways that turn her partly into a kind of geometric figure in which her face is squared off, in which her various body parts, insofar as they're represented at all, are represented in forms that systematically distort and have a kind of mechanizing or abstracting or geometric element or dimension.

Another way to think about this distorting element is to think about the satiric or disgusted images of depravity and nightmare that we find in German Expressionism, another aspect of painterly Modernism, but influenced especially by certain tendencies within German culture itself. Franz Kafka, the great writer we'll be looking at later in these lectures, is a part of and very influential figure in this movement of Expressionism.

For our purposes now, though, we might look at a fairly famous painting by the Expressionist painter Max Beckmann. There are a number of paintings that he did that would fit the bill, but there's one especially called *Family Picture* that I want to talk about briefly. It now exists in a New York museum.

I want to describe this last unforgettable expression of thoughtful contempt and disgust at the human family. The painting shows a strange grouping of human beings in a room. The perspective is altered in the painting as if in a dream. The characters are distorted and unrealistically represented in some ways, although they're perfectly recognizable as figures.

The space is cramped and confined, a single room in which the ceiling and the floor seem to be pressing the characters together, and there's a horizontal pressure against them as well. They're sort of crammed together across the horizontal length of the picture. There's a child sprawled like a small octopus in the lower foreground. The heads of all the figures are much larger than the bodies.

There's a standing woman in a red slip, her cone of red hair literally hitting against the ceiling, or maybe you have the feeling the ceiling is pressing down on her hair, and she's admiring herself in a mirror. Seated beside her is a woman I'll call grandma, an older woman. Grandma covers her face in woe like the character in Munch's *Scream*. There's a dwarfish gross man with a pipe and leprechaun-type shoes reclining next to the standing woman. They look like ugly characters from a nastier and meaner Maurice Sendak.

None of the figures, crushed together as they are, looks at one another or seems aware of anyone. Garish colors, mushroom-shaped lights coming out of the ceiling, further distort the atmosphere. Beckman is illustrating *The Metamorphosis* or being inspired by *The Metamorphosis*, and it's not an accident that the editor and translator

Stanley Corngold chose to reproduce this picture on the cover of his edition of *The Metamorphosis*.

But I'd like to end on a less disgusted note in talking about these paintings with an earlier and centrally Postimpressionist painting, which I think speaks more directly, more fully, more completely, and maybe more generously to the idea of Modernism that our course is interested in. It's an immensely greater painting as well. It's van Gogh's *Starry Night*, painted in 1889.

Remember this wonderful painting? A rural village is seen from a hilltop or a high vantage point. On the left of the painting there's a fiery evergreen, a cypress, thrusts flame-like up into the sky to the top of the painting.

Down lower in the valley, in the very center of the canvas, there's a church steeple, but the church steeple is dwarfed in a way by the fiery, flaming, thrusting energy of the "tree" on the left, and also by the swirling, roiling electric strokes of blue and white that seem to show the sky in a vibrating motion; and the vibrating motion of the sky is at the very center of one's experience of the picture. No conventional stars are to be seen, but balls of swirling yellow and white and orange, almost as if miniature novas, miniature supernovas are in the middle of the painting.

And when you're in the presence of the painting, no reproduction does it full justice. When you're actually in the presence of the painting, you become aware of the thickness of the paint. You want to reach out and touch it. If there weren't guards there to keep you from doing so, many people would probably do so, because the thickness of the paint is so much an aspect of your experience of the painting. It makes you aware of the painting as a painting, as an object made of oil paints and canvas, and it also suggests a kind of immense self-reflexiveness or self-awareness on the part of the painter.

It's a picture, in a way, about the making of the picture. There's an emotional dimension to it. You feel that the painter felt a kind of desperation to capture something that's disappearing in the very instant that he's seeing it, in the very instant of its occurring; as if he's trying to capture the uniqueness and the fluidity of this single moment. So, you feel the evanescence of the moment but also its incredible beauty; and part of the reason is that you feel that van

Gogh is in a kind of desperate frenzy himself to capture it before it disappears.

We know, in fact, that he put some of the paint on the painting with his finger, with his thumb, and you can see thumbprints on the painting as well. So that's another way in which the subjectivity, the personality of the artist is a part of your experience of the painting. The emotional dimension of the painting, the feeling you get when you look at the painting, is something of van Gogh's own desperation and own desire to represent the world and also his fear that he can't do it; that the world is escaping him even as he's creating the painting is a profound aspect of one's experience of looking at this remarkable work of art.

The agony and the ecstasy of this astonishing painting will be replicated, though in verbal registers, in Conrad, in Ford Madox Ford, in Joyce, and maybe most compellingly in D.H. Lawrence and Virginia Woolf.

But to conclude this introduction to Postimpressionism and as a context for our literary study, I'd like to return to literature proper, and to begin to give an answer to the first and much more complex question: What did Virginia Woolf mean when she said that human character had changed? I think part of the answer, but only part of it (the full answer will emerge, I think, through the whole course of these lectures), part of the answer can be found in a simple exercise or teaching strategy that I learned from Richard Ellmann from his Modern fiction course many years ago and have been deploying with my own variations ever since. It's another way to distill elements of the Modern as part of a context for our reading of particular texts.

In my classes on Modern fiction I usually distribute a series of first paragraphs to 19th- and 20th-century books without identifying them, to whet the students' appetite for reading and for the adventure of our course. And I want to try a more modest version of that here, so I'll read first without identifying the paragraphs, but I won't leave you in suspense very long. As soon as I complete the paragraph I will tell you who wrote it.

And what I'd like to do is to compare again the earlier with the later paragraphs. They're all beginnings. They're all the very beginnings of novels. And I think doing so will help to clarify some of the profound differences between these two traditions of literature, and will help us to understand what is unique and distinctive about the Modern. Beginning no. 1:

> Emma Woodhouse, handsome, clever, and rich, with a comfortable home and happy disposition, seemed to unite some of the best blessings of existence; and had lived nearly twenty-one years in the world with very little to distress or vex her.

Well, of course, that's Jane Austen, *Emma,* 1815. One of the things to note about the paragraph, of course, is how confident Austen is about the shared vocabulary of moral understanding she has with her audience and also, of course, the expository clarity of this beginning. The reader is situated instantly. Some fundamental things about Emma's nature are generalized in a confident omniscient way by Austen's voice. Emma's clever, she's rich, she's lived comfortably, she has a happy disposition, she unites the best blessings of existence. Well, Austen has confidence that a phrase like "best blessings of existence" will signify for her audience.

Now let's compare that with a 20th-century beginning:

> He was an inch, perhaps two, under six feet, powerfully built, and he advanced straight at you with a slight stoop of the shoulders, head forward, and a fixed from-under stare which made you think of a charging bull. His voice was deep, loud, and his manner displayed a kind of dogged self-assertion which had nothing aggressive in it. It seemed a necessity, and it was directed apparently as much at himself as at anybody else. He was spotlessly neat, apparelled in immaculate white from shoes to hat, and in the various Eastern ports where he got his living as a ship-chandler's water-clerk he was very popular.

Well, maybe the ship-chandler gives away the paragraph. It's a paragraph by Conrad. It's the beginning to his great novel, *Lord Jim,* published in 1900. What a contrast with Austen. Think about it for a moment. Conrad's narrator doesn't even have the confidence to tell us how tall this man is. He's an inch or perhaps two under six feet. Well, which is it? There's a tentativeness, an uncertainty, a qualified fearfulness, almost, about the narrator's willingness to generalize.

Listen again: He had "a fixed from-under stare which made you think of a charging bull." And he says, "His manner displayed a kind of ... self-assertion which ... seemed a necessity." Well, doesn't the narrator know? And of the course the answer is, he doesn't. "It was directed apparently as much at himself ..." The sense of uncertainty, the sense of tentativeness is profound.

But there's something else. How little real exposition is there? When we finish that Conrad paragraph, we know something about Jim, a good deal in fact, and many of those details will become much more resonant as the book goes on, but we certainly are not situated clearly in a morally coherent universe whose basic parameters have already been established by a confident, all-knowing author.

Let's try another example—19th-century beginning:

> While the present century was in its teens, and on one sunshiny morning in June, there drove up to the great iron gate of Miss Pinkerton's academy for young ladies, on Chiswick Mall, a large family coach, with two fat horses in blazing harness, driven by a fat coachman in a three-cornered hat and wig, at the rate of four miles an hour. A black servant, who reposed on the box beside the fat coachman, uncurled his bandy legs as soon as the equipage drew up opposite Miss Pinkerton's shining brass plate, and as he pulled the bell at least a score of young heads were seen peering out of the narrow windows of the stately old brick house. Nay, the acute observer [19th-century fiction is full of acute observers] Nay, the acute observer might have recognized the little red nose of good-natured Miss Jemima Pinkerton herself, rising over some geranium pots in the window of that lady's own drawing-room.

That's Thackeray from his great novel *Vanity Fair*, published 1847–1848. What an outstanding and remarkable and very beautiful paragraph. I don't mean this comparison to suggest that 19th-century fiction does not give us immense pleasures. It does. It's a very great, very rich fiction. I'm just trying to describe fundamental differences between the two traditions.

In this paragraph, think for a moment of how much this narrator claims to know, and how detailed his knowledge is. The present century is in its teens. The large family coach has not just horses; they are fat horses. There's a fat coachman. He knows what his clothing is like. He even knows how fast the coach is going—four miles an hour. There's a precision here that suggests a kind of knowledgeability that has no end, as if the omniscient powers of the narrator are so great that there's no detail unavailable to his hovering intelligence, to his omniscient understanding.

And when we get to the passage that talks about "at least a score of young heads" looks out, this omniscient perspective also has the power to see from many different angles at once. It can focus down closely if it wishes to, or it can pull back in a kind of panoramic view and give us a whole panorama. The immense power, the immense confidence of this narrating intelligence is one of the delights of the paragraph, and one source both of its verbal energy and of its comedy.

Well, compare this beginning, this first paragraph:

> Through the fence, between the curling flowers flower spaces, I could see them hitting. They were coming toward where the flag was and I went along the fence. Luster was hunting in the grass by the flower tree. They took the flag out, and they were hitting. Then they put the flag back and they went to the table, and he hit and the other hit. Then they went on, and I went along the fence. Luster came away from the flower tree and we went along the fence and they stopped and we stopped and I looked through the fence while Luster was hunting in the grass.

Now, my goodness. What is going on here? If you've never read this book, you have no idea; you're lost; you're completely baffled. Who's speaking here? What is he describing? And of course, that is the beginning of Faulkner's great novel, *The Sound and the Fury*.

It's the voice of the idiot Benjy, who is the narrator of the first section of the novel. And you have to read through that entire first section before you even begin to get some minimal sense of what Benjy is talking about. Benjy doesn't have a language or a vocabulary or an understanding to recognize that, of course, what he's describing is a game of golf. And we come to realize that later in the novel.

He's running along the fence because as we learn, again, much later in the novel, his favorite playground, his favorite play area was this wonderful pasture which his father sold in order to get money to send one of his sons, Quentin Compson, to Harvard. So he sold the pasture off, and poor Benjy no longer is allowed to go into the pasture, so he runs along the side of the pasture and looks over the fence at the golfers.

The pasture has now become a golf course. Luster is the black servant who takes care of Benjy, and when Luster is hunting in the grass, he's hunting for golf balls. Now, none of these facts become available to us until we're much deeper into the novel—50 or 75 or 100 pages in before all of these details are clear to us.

Faulkner is obviously up to a very different game. He plunges us in the middle of things, into a consciousness whose nature we need to decipher, whose nature we need to understand. And he doesn't speak at all with the all-knowing, confident knowledgebility of a 19th-century omniscient narrator. Modernist writers, Modernist narrators—not just first-person narrators, but even Modernist narrators who claim something of the third-person powers of the old omniscient narrators of the 18th and 19th centuries—have radically reduced powers. Why?

Part of the reason is that they live in this skeptical, pessimistic, problematic, disorderly environment I described in our previous lectures. They are the heirs to uncertainty. They are the heirs to instability. They are the heirs to a notion of the fluidity and immense complexity of a world that no longer has the kind of stabilizing coherences that earlier religious and moral and institutional dispensations created for people. And the beginnings of Modern novels, therefore, reflect this much more problematic, this much more fluid, this much more uncertain environment.

One more example:

> With a single drop of ink for a mirror, the Egyptian sorcerer undertakes to reveal to any chance comer far-reaching visions of the past. This is what I undertake to do for you, reader. With this single drop of ink at the end of my pen, I will show you the roomy workshop of Mr. Jonathan Burge, carpenter and builder, in the village of Hayslope, as it appeared on the eighteenth of June, in the year of our Lord 1799.

George Eliot, *Adam Bede*; one of the great writers of all time. Many things to note here, but the confidence of the author—she can even call attention to her writerly process without threatening or without feeling that she's threatening the plausibility or believability of her world. "With this single drop of ink," reader, I will recreate the world.

Listen to this contrasting passage:

> To begin with I wish to disclaim the possession of those high gifts of imagination and expression which would have enabled my pen to create for the reader the personality of the man who called himself, after the Russian custom, Cyril son of Isador ... Razumov.

And of course that is the beginning of Conrad's great political novel, *Under Western Eyes*. Compare the confident self-consciousness of George Eliot to the nervous, apologetic, "I can't write," confessional quality of Conrad's first-person narrator.

Henry James, one of the great theorists of Modern fiction, spoke of Modern fiction as being committed to what he called the "muted majesty of authorship." The majesty of authorship in Modern fiction is muted because the Modern writers are much more conscious than their predecessors of the limitations of language, the limitations of our conceptual categories, the limitations of words and of artistic traditions themselves.

In the problematic and uncertain universe of Modern fiction there are other compensations and other virtues, but the certainty and the self-confidence of the older tradition is gone.

Lecture Five
The Man Who Would Be King—Imperial Fools

Scope:

Immensely famous in his own lifetime, this Nobel Prize winner's reputation has suffered a radical decline, in part because of his imperialist politics. Kipling was a gifted, perhaps immortal children's writer (*The Jungle Book*). His colloquial poetry—often in the voices of working-class British soldiers who are themselves victims of the imperium they serve—was learnt by heart by generations of British and American school children. We'll sample and briefly discuss Kipling's poems before examining one of his most famous stories, *The Man Who Would Be King*. We'll analyze the opening paragraphs of this novella closely, watching for the irony in the first narrator's description of "loafers," "natives," and capitalized "Deficits." Then we'll examine the transition to the colloquial primary narrator, one of Kipling's richest characters, and consider the implications of his apparent unreliability, and the larger ambiguities in Kipling's vision of empire. Is this parable of imperial over-reaching and catastrophe still relevant today? Many believe so, and not only because the adventure takes place in Afghanistan.

Outline

I. We begin with a general sketch of Kipling's life and career.
   A. Kipling (1865–1936) was a transitional figure in Modern literature. The first Englishman to win a Nobel Prize for Literature (1907), he was immensely famous in his own lifetime.
   B. His reputation has suffered a radical decline in recent years, in part because of his imperialist politics.
      1. He visited South Africa often in the early 20th century, when he was already one of the most famous writers in the English-speaking world.
      2. He associated with the British-born Cecil Rhodes, the well-known racist and conqueror of southern Africa,

whom many thought had influenced Kipling's imperial views.
   3. The racism and jingoism in Kipling's work have somewhat obscured his genius, but his most compelling and influential work transcends such reductiveness and inhumanity.
   4. His infamous poem "The White Man's Burden," written in 1899 as advice to the United States, embodies the West's racist attitudes toward Africa and Asia at that time.

II. Kipling wrote in a variety of genres.
   A. He was a gifted writer of children's books.
      1. His most well-known works include *The Jungle Book* (1894), *The Second Jungle Book* (1895), and *Just So Stories* (1902).
      2. Even these contain authoritarian and perhaps racist undercurrents, however.
   B. He also wrote *demotic*, or conversational, poetry, the most famous of which was "Gunga Din."
   C. Some of Kipling's other notable works include fiction and poetry.
      1. *Departmental Ditties* (1886) and *Barrack-Room Ballads* (1892) are collections of poems.
      2. *Captains Courageous* (1897) and *Kim* (1901) are novels.

III. Kipling's most powerful art resides in a dozen or so stories and novellas, chief among which is the novella, *The Man Who Would Be King* (1888).
   A. We can establish a context for Kipling and, specifically, for this novella and the traditions against which it argues by examining the deeply embedded attitudes about Europe's imperial destiny that dominated the 19th century.
   B. One person who embodied the myth of imperial destiny was Sir James Brooke (1803–1868), known as the Raja of Sarawak (in Borneo).
      1. Brooke served in the East India Company Army in Burma until 1830. This private, imperial corporate army was dissolved following a Bengali uprising in 1857,

after which control of India's government was transferred to the British Crown.

    2. This event is referred to at the end of *The Man Who Would Be King* by the character Peachey Carnehan, who says to his comrade, Dan Dravot, "I'm sorry, Dan.... This business is our Fifty-Seven."

C. Later, Brooke went to Borneo and helped local officials put down a tribal rebellion. He was made raja in 1841, after which he governed the country, devised a tax system, and personally administered justice—just as Dravot does in Kipling's story.

D. Brooke is explicitly referred to in Kipling's work, and Dravot and Carnehan model him knowingly, illustrating the power of the myth of white supremacy and implying that native cultures welcomed "enlightened rule" by imperialists.

E. Popular culture of the time also mirrored this imperial myth, as is evident in the 1909 music-hall song titled "I've Got Rings on My Fingers."

    1. This ditty tells of the Irishman Jim O'Shea who traveled to an Indian island and became the boss simply because of his Irish smile.

    2. The song even uses native words, such as *nabob* and *panjandrum*, which became part of the English lexicon.

    3. The song, then, is both attentive to native culture and condescending toward it.

IV. Let us look more closely at the story of *The Man Who Would Be King*.

A. This work retells the tale of the white conqueror but with a deconstructive, ironic energy.

B. The story is told by an unnamed first-person narrator who is a newspaper editor. He frames the story of two former British soldiers freelancing as blackmailers and charlatans in colonial India.

C. The two soldiers, Dan Dravot and Peachey Carnehan, undertake the mad trek to "Kafiristan," a section of Afghanistan that they intend to conquer. Three years later, looking haunted and terrible, Peachey recounts the story of their adventures to the narrator after his return to England.

D. The opening paragraphs contain an ironic, generalizing, 19th-century voice that ultimately exposes the casual, inhumane racism of the "proper" British colonial.
   1. The opening of the story illustrates a system of class hierarchies in which races and classes are separated in train carriages.
   2. Implicit in this opening is a profound critique of the imperial attitude and the imperial adventure overall.

E. The heart of the novella is Peachey's brilliantly vivid and colloquial narration.
   1. His account comes alive, not with *ex cathedra* pronouncements but with concrete details, the compelling ambiguities of actuality.
   2. He is a powerful example of a damaged or unreliable narrator.

F. Peachey recounts the killing of Dravot, which results from Dravot's Faustian overreaching.
   1. He and Peachey had succeeded in becoming kings of Kafiristan, but in the story's richest irony, they are brought low because Dravot abandons his original impulse to loot the country and, instead, decides to join his kingdom to the British Empire.
   2. Perhaps the most poignant and disturbing irony here is that the Scotsman Dravot and the Irishman Carnehan are themselves part of an oppressed underclass whose own countries fell under British rule long before they signed on as soldiers of Victoria's empire.
   3. These deluded, brutal adventurers have internalized the values of their conqueror and are destroyed by them, making this a powerful anti-imperial fable.

G. But some contradictions keep the tale from reaching the highest level of art, and these surely stem from Kipling's own divided nature.
   1. The emphasis at the end is on the horrific detail of Peachey having carried Dravot's severed head in a sack on his stumbling journey back from Kafiristan to England.
   2. This climax displaces the reader's attention from the story's critique of imperialism. Further, the severed

head, and Peachey's crucifixion both make the natives look more barbaric than their conquerors.
H. Damaged or disappointing endings are common in great stories, however. Despite its faltering conclusion, *The Man Who Would Be King* remains a landmark of English literature, as disturbing and instructive today as it was nearly 120 years ago when it was first published.

Essential Reading:
Kipling, *The Man Who Would Be King*.

Supplementary Reading:
Kipling, *Kim*.

Questions to Consider:
1. How does the colloquial voice of the primary narrator in *The Man Who Would Be King* differ from that of the newspaper editor who "frames" the story?
2. What is implied by the fact that Dravot and Carnehan, the "kings" of the story, are Scottish and Irish, that is, they are natives of countries like India that are part of Britain's empire?

Lecture Five—Transcript
The Man Who Would Be King—Imperial Fools

We begin our adventure in Modern fiction with a transitional figure, Rudyard Kipling. His dates are 1865 to 1936, and although he outlived a number of the writers we'll be studying, he really belongs in a certain sense at least as much in the 19th century as in the 20th, and he's instructive exactly because of this transitional quality.

Immensely famous in his own lifetime, the first Englishman to win a Nobel Prize for literature, in 1907, his reputation has suffered a radical decline in recent years, in part because of his imperialist politics. He visited South Africa often in the first decade of the 20th century (he was already, of course, a famous man, one of the most famous writers in English) and he was given a house to live in by Cecil Rhodes, the famous philanthropist/racist/ imperial conqueror whose name is given to the Rhodes Scholarship, and whose name had been noted in the country called Rhodesia (now Zimbabwe). Rhodes loaned Kipling a house to live in on a number of Kipling's visits to South Africa, and that association with Rhodes was said by many people to have fostered Kipling's imperial politics, his attitudes toward British imperialism.

The most infamous expression of this in Kipling's work is a poem, an infamous poem entitled "The White Man's Burden," and you can get some sense of how extreme these views were and how widespread they were, I suppose, from even the very first stanza. This poem was first published in February 1899 in the *London Times*, and it was addressed to the United States. He was flattering us I suppose in a way. It came in the wake of the American war with Spain in 1898, which led to the acquisition of Cuba and the Philippines, which had been previously Spanish colonies. Kipling sent an advance copy of this poem to his dear friend Theodore Roosevelt.

This is the beginning of the poem:

> Take up the White Man's burden—
>   Send forth the best ye breed—
> Go, bind your sons to exile
>   To serve your captives' need;
> To wait, in heavy harness,
>   On fluttered folk and wild—

> Your new-caught sullen peoples,
> Half devil and half child.

Well, the unbelievable explicit racist condescension there is alas characteristic of certain European attitudes toward Asia and Africa all throughout the 19th century. And it's important to keep these attitudes in mind, in part because they make Kipling's great works—which question such attitudes powerfully and profoundly—even more remarkable, as if he was writing against the grain of his own instincts sometimes in his very best work.

Kipling was born in Bombay to British parents. His father, John Lockwood Kipling, was a school principal and museum curator, and the illustrator of a book that was very widely circulated in English-speaking countries titled *Beast and Man in India*.

Kipling, as many people know, was a gifted, perhaps immortal children's writer. His most famous children's books are *The Jungle Book* and *The Second Jungle Book,* published in 1894 and 1895; and a wonderful set of stories that I still remember from my childhood called *Just So Stories*, which tell a series of animal fables.

One of my favorite of these animal fables (this is sort of a self-indulgence, but it's such a nice story that I can't resist mentioning it to you) is the tale "How the Rhinoceros Got His Skin," and it tells the story: In the olden days the rhinoceros's skin was very smooth. He wore it like a coat and it had buttons on it, and he would button it up. And it was perfectly smooth.

One day the rhinoceros went down to the river to go for a swim and he unbuttoned his outer skin and left it on the bank of the river while he went swimming; and while he was there a character called a Parsee-man (Kipling's condescending treatment of a Hindu religious figure) is angry at the rhinoceros because he had eaten some of his specially baked cakes, and he sneaks down to the river while the rhinoceros is swimming and he drops some bread crumbs into his skin.

When the rhinoceros comes back and puts the skin back on, his back itches, his body itches like crazy, so he rubs himself against all manner of things, including a very jagged palm tree, and rubs the buttons of his skin off, so it's stuck forever. And the skin becomes more and more wrinkled. So, that's how the rhinoceros got his skin.

And the moral of the story tells us sadly how even Kipling's racist attitudes got snuck into his children's stories, went like this: "Them that takes cakes that the Parsee-man bakes makes dreadful mistakes!"

I remember as a child thinking that was incredibly witty and loving the comical rhymes in the line without understanding, of course, that the mockery of inadequate English and the reference to the Parsee man was in fundamental ways condescending. But his children's stories are, I think, in many ways, despite this limitation, very remarkable. Both *Jungle Books* contain much richer and more wonderful stories than their Disney versions.

He was also famous in his own lifetime for what we might call his conversational or demotic poetry, often written in the voices of working-class British soldiers. Some of these poems were learned by heart by generations of British and American school children. I suppose the most famous of these, although there are many other wonderfully well-known ones, is the poem "Gunga Din." And I want to read you a couple of fragments of "Gunga Din" to give you a sense of Kipling's gifts as a conversational poet. His poetry was not dense or deep or difficult, but it captured ordinary British speech in a quite remarkable way, and especially the British speech of the uneducated, the soldier class. Some of his speakers use cockney accents, for example, and it's very interesting. You can understand why the poetry was so successful, because it appealed so broadly across various social classes.

Here are some passages from "Gunga Din":

> Now in Injia's sunny clime,
> Where I used to spend my time
> A-servin' of 'Er Majesty the Queen,
> Of all them blackfaced crew
> The finest man I knew
> Was our regimental bhisti, Gunga Din. [*Bhisti* means water carrier.]

And here are a few passages from later in the poem:

> I shan't forgit the night
> When I dropped be'ind the fight
> With a bullet where my belt-plate should 'a' been.
> I was chokin' mad with thirst,
> An' the man that spied me first

> Was our good old grinnin', gruntin' Gunga Din.
> 'E lifted up my 'ead,
> An' he plugged me where I bled,
> An' 'e guv me 'arf-a-pint o' water-green:
> It was crawlin' and it stunk,
> But of all the drinks I've drunk,
> I'm gratefullest to one from Gunga Din.
>
> 'E carried me away
> To where a dooli lay, [A *dooli* is a stretcher.]
> An' a bullet come an' drilled the beggar clean.
> 'E put me safe inside,
> An' just before 'e died,
> "I 'ope you liked your drink", sez Gunga Din.
> So I'll meet 'im later on
> At the place where 'e is gone—
> Where it's always double drill and no canteen;
> 'E'll be squattin' on the coals
> Givin' drink to poor damned souls,
> An' I'll get a swig in hell from Gunga Din!

Well, you can see in some ways it's an irresistible poem despite the residue of condescension even in this poem. At the poem's grand finale, Kipling says:

> Though I've belted you and flayed you,
> By the livin' Gawd that made you,
> You're a better man than I am, Gunga Din!

And I suppose at some level the poem means it, the speaker in the poem means it, but we can also feel some level of condescension and some residue of racial incomprehension in the poem.

Some of Kipling's major titles are perhaps worth mentioning. These conversational poems I've been referring to were collected in two volumes that are still in print: *Departmental Ditties*, published in 1886, and *Barrack-room Ballads*, 1892, which is the book that contains "Gunga Din."

He also, in 1888, collected a first group of wonderful stories mostly set in India—and in other exotic places—especially India, and one of the stories in that volume was *The Man Who Would Be King*, the text we'll be looking at in a moment.

He published some novels as well. The two most well-known novels are a short novel called *Captains Courageous* in 1897, a coming-of-age novel about a rich boy who was knocked overboard off a luxury liner and is picked up by a working vessel, and they won't take him back to shore, so he has to spend whatever it is—several months—aboard this fishing boat, and he becomes a man because he's surrounded by ordinary working men, and he has all of his upper class pretensions burned out of him in the novel.

Probably his best novel, and his best-known novel, published in 1901, is a novel entitled *Kim*. It's set in British India and it follows the exploits of a young boy who, among other things, is brought into the British Secret Service and does various spying activities in what Kipling calls "the great game," the conflict between Britain and especially Russia over control of the Asian subcontinent.

The racism and jingoism in Kipling's work has obscured his genius somewhat, but his most compelling and influential work transcends such reductiveness and inhumanity. His most famous novels, as I said, show a side of his genius, and there are other works, especially short stories, which show it as well. But there's a consensus that his most powerful art resides in a series of stories and novellas of which the most important and the most well known is probably *The Man Who Would Be King*.

You may recall that there was a very interesting film by John Huston in 1975 made of the story. It starred Michael Caine as Peachy Carnehan and Sean Connery as Daniel Dravot, the two adventure partners in the story. And it starred Christopher Plummer as the outer narrator, and the Plummer character was made up to look exactly like photographs of Kipling, and I think he was called "Kipling" in the film. And there was no question you were supposed to think of him as Rudyard Kipling. There's no harm in our imagining that the first narrator, the outer narrator of *The Man Who Would Be King* is Kipling as well, even though he's not so identified.

One way to establish a context for *The Man Who Would Be King*, to clarify its importance and the intellectual and psychological traditions against which it is arguing, which it is in some sense exposing, is to talk about attitudes toward Europe's imperial destiny that were dominant at the time that *The Man Who Would Be King* was published. Those attitudes are implicit already in some of the poetry that I've quoted to you from Kipling.

But there are certain other factors that I think will also help to make these attitudes clear. One of the most important of these contexts is the context for establishing what might be called the cultural myth of Europe's imperial destiny, and to understand how deeply that myth was embedded in the DNA of European culture throughout the 19th century, is to talk about a real-life character, a man who was knighted for his activities, Sir James Brooke. His dates are 1803 to 1868, and he was known at the time of his death as The Rajah of Sarawak, a province or a country in Borneo—that's a large island in southeast Asia that today comprises Malaysia, Indonesia, and Brunei.

James Brooke was born, like Kipling, in India of British parents, and he served first in the East India Company Army in Burma. Interesting in itself is the fact that this army was a private army, a corporate army, run by the East India Company. This imperial corporate army was dissolved after an uprising by a Bengali native army in 1857, which led to the transfer of the government of India from control by the East India Company (which was a trading company) to the British crown in 1858.

I mention this event because it's actually referred to at the end of *The Man Who Would Be King*. At a certain point, just as they're near their downfall, Peachey Carnehan turns to Dan and he says: "I'm sorry, Dan ... This business is our Fifty-Seven," and he's referring to the 1857 uprising by that Bengali army. So the imperial history of British involvement in India and Asia more generally is embedded in explicit references in *The Man Who Would Be King*.

After that experience, Sir James Brooke went to Borneo where he helped a local official put down a tribal rebellion. He left that corporate army and in effect he traveled as a private mercenary to Borneo, became involved with a tribal chief, helped put down a rebellion, and he was made a rajah (a ruler) in 1841, after which he governed the country, devised a tax system, personally administered justice, just like Dravot in *The Man Who Would Be King*. His story was very widely publicized, and he was knighted for his activities. His descendants remained in Borneo until the Japanese conquest in 1942. A biography of him published in 1960 is titled *The White Rajah*.

And Brooke is explicitly referred to in Kipling's novella; Dravot and Carnehan literally model him. They think of him as their ancestor or as the model they're following, and at a certain moment in the story, Dravot cries out—this is at the zenith of his kingly and dynastic delusions—he cries out: "Rajah Brooke will be a suckling to us," he says.

So I mention this story not only because it eliminates certain somewhat obscure references in *The Man Who Would Be King*, but also because it tells us something about how powerful this myth of white supremacy was. The idea was James Brooke went to these benighted territories and brought them a kind of enlightened rule. And the implication is also that they liked this; they welcomed this.

But I can give you an even more dramatic illustration of how powerful this myth was to Europeans at the time, a myth that *The Man Who Would Be King* explicitly undermines and deconstructs, by referring to a famous music hall song. It's dated 1909, but it surely represents attitudes and feelings that are a century old. Forgive my bad singing, but I will try to partly sing it so you'll get a feeling for what the song is like. It's a song some of you may have heard of. The title is *I've Got Rings on My Fingers*, and here's how it begins:

> Jim O'Shea was cast away
> Upon an Indian isle
> The natives there they liked his hair
> They liked his Irish smile
> So made him king Panjandrum
> *The Nabob of them all*
> They called him Jiggi-Bob-Jay
> And rigged him out so gay
> That he wrote to Dublin Bay
> To his sweetheart, just to say
>
> Sure I've got rings on my fingers, bells on my toes
> Elephants to ride upon, my little Irish Rose
> Come to your Nabob, and on next Patrick's Day
> Be Mistress Mumbo Jumbo Jiggi-Bob-Jay O'Shea.

Well, there are more stanzas, but you don't need to hear more of my singing. I mention the song as a way of showing you how deeply embedded these imperial attitudes were, because here this is part of the popular culture, as if every school child might be singing such songs. But the fact that such mythologies are so deeply embedded in the popular as well as the learned consciousness of the society seems very important.

And the song is interesting in a number of ways. One of the ways it's interesting is that it explicitly enacts this myth of European superiority. Jim O'Shea goes there. They like his Irish smile, so they make him the boss. What a fantasy of European superiority is embedded there! And there's also at the same time a recognition, I think, in the song's use of the native words like *panjandrum* and *nabob*, of how widespread these native terms had become in the English language by the end of the 19th century.

A white man's fantasy, then, the song of the complacent, self-deluding conqueror. Well, the novella to which we will now turn articulates this meaning very powerfully, very fully. *The Man Who Would Be King* retells this tale, but with a deconstructive ironic energy. And, of course, the novella has special relevance in today's world, because it's set, as you may know, in Afghanistan, and among the imperial adventures of a Western society trying to control Afghanistan. Afghanistan is back in the news, as you may know, and I saw a recent article in the *New York Times* by Edward Rothstein in 2002. It was titled "Kipling Knew What the United States May Now Learn." And it might well be that *The Man Who Would Be King* would be very good reading for policymakers in the United States.

The basic plot of the story: An unnamed "outer" first-person narrator (really Kipling), a sometime spy, a newspaper editor, frames the story of two former British soldiers, now freelancing as blackmailers and charlatans in colonial India. After consulting maps and drawing up a "Contrack"—that's a quotation—which includes abstaining from drink and women, Dravot and Carnehan undertake a mad trek to "Kafiristan"—a province or section in "the top right-hand corner of Afghanistan," the text tells us, which they intend to conquer with their soldier's knowledge of drill formations and rifles.

They make their contract in the newspaper office in the presence of this outer narrator, and then three years elapse; and then Peachey Carnehan returns, looking haunted and terrible, a bit crazy, to the newspaper office, where he recounts his adventures, again to the framing outer narrator.

In the opening paragraphs of the story we can hear something of the generalizing 19th-century voice of an omniscient narrator that we've already noted in those first paragraphs we looked at, and this is the Kipling voice (the outer narrator):

> The beginning of everything was in a railway train upon the road to Mhow from Ajmir. There had been a Deficit in the Budget, [Kipling capitalizes D and B—Deficit and Budget—creating a kind of irony for an attentive reader, as if he's sort of mocking these qualities—the people who take Budgets and Deficits seriously] which necessitated traveling, not Second-class, which is only half as dear as First-class, but by Intermediate, which is very awful indeed. There are no cushions in the Intermediate class, and the population are either Intermediate, which is Eurasian, or native, which for a long night journey is nasty, or Loafer, which is amusing though intoxicated. Intermediates do not patronize refreshment-rooms. [So far the paragraph is full of certain comic surprises, and maybe we can hear a hint of condescension in what's being described, but still we might not fully realize what's at issue until this sentence.] They carry their food in bundles and pots, and buy sweets from the native sweetmeat-sellers, and drink the roadside water. That is why in the hot weather Intermediates are taken out of the carriages dead, and in all weathers are most properly looked down upon.

A brilliant ironic paragraph I hope you recognize in which a crescendo of details ultimately exposes the casual inhumane racism of the "proper" British colonial. The surprise and moral horror of the final words still get me after multiple readings. "That is why in the hot weather Intermediates are taken out of the carriages dead, and in all weathers are most properly looked down upon."

One of the things this paragraph shows us is that the system is striated with class hierarchy. There are three classes of transport, a segregating of races and classes, and a rail system so efficient (as we learn a few paragraphs later) that Peachey Carnehan can persuade the narrator that he will be able to deliver a message to Dravot because "he'll be coming through Marwar Junction in the early morning of the 24th by the Bombay Mail" (the Bombay Mail is a train).

Now think about that. The rail system is so accurate, so reliable that you can absolutely predict to the moment, to the hour, perhaps even to the minute, when particular trains will converge. And it's a system with all these different classes of transport, and yet it's also a system in which the roadside water is so dangerous that it kills the natives. Implicit in the paragraph itself is a profound critique of the imperial attitude, of the imperial adventure more broadly.

My students often falter over the embedded ironies in the paragraph I've just read to you. They're not sure whether the Kipling-narrator endorses the inhuman snobbery of the paragraph. But I think that the capital letters in Deficit and Budget imply this, and of course it is not "most proper" to be oblivious to such inequities and suffering. The paragraph is a powerful indictment of imperialism.

But the heart of the novella, the source of its artistic excellence and moral power, derives from Peachey's brilliant, vivid, colloquial narration. Once the story shifts to Peachey's account, it comes alive in a new way, in a powerful and partly Modern way—not with ex cathedra pronouncements like those just quoted from the beginning of the story, but with concrete details, the compelling ambiguities of actuality.

Peachey is a powerful, even pioneering, example of a damaged or unreliable narrator. Listen to his half-mad voice as he describes Dravot's death:

> They marched him a mile across that snow to a rope-bridge over a ravine with a river at the bottom … They prodded him behind like an ox. 'Damn your eyes!' says the King. 'D'you suppose I can't die like a gentleman?' [The king, of course, is Dravot.]

> … Out he goes, looking neither right nor left, and when he was plumb in the middle of those dizzy dancing ropes, 'Cut, you beggars,' he shouts; and they cut, and old Dan fell, turning round and round and round, twenty thousand miles, for he took half an hour to fall till he struck the water, and I could see his body caught on a rock with the gold crown close beside.

This catastrophe results, of course, from Dravot's Faustian overreaching. He and Peachey had succeeded in becoming kings of Kafiristan, taken by the natives for gods really. But in the story's richest irony, they are brought low because Dravot abandons or transcends his original impulse just to loot the country for treasure.

Peachey was criticizing him for this. Peachey just wanted to take the money and run. But Dravot, ironically, comically, mordantly, terribly in the end, decides that he has larger ambitions.

He wants to take a queen. He wants to establish a dynasty, and he wants to join his kingdom to the British Empire: "I won't make a Nation," he says, "I'll make an Empire! These men aren't niggers;" he says, "they're English! Look at their eyes—look at their mouths … They sit on chairs in their own houses … They only want rifles and a little drilling. Two hundred and fifty thousand men, ready to cut in on Russia's right flank when she tries for India! … When everything was ship-shape, I'd hand over the crown … to Queen Victoria on my knees, and she'd say, 'Rise up, Sir Daniel Dravot.'"

Perhaps the most poignant and disturbing irony here is this: Dravot the Scotsman and Carnehan, an Irishman, are themselves part of an oppressed underclass, whose own countries fell under British rule long before they signed on as soldiers of Victoria's empire! Like their native sidekick, Billy Fish, in the story, who dies by choice with Dravot at the end instead of melting back into the native population, these deluded, courageous, and brutal adventurers have internalized the values of their conqueror and have been destroyed by them.

It's a powerful, powerful anti-imperial fable. But there are some contradictions in it that perhaps keep the tale from reaching the highest level of art, despite its many astonishing virtues. Some of the contradictions and confusions, both in the story itself, surely come from Kipling's own divided nature. In the end, for example, of the story, there's an emphasis on a horrific detail: Peachey has carried a sack with him on his deranged return trek across desert and mountain, and he opens the sack as he's narrating this tale to the Kipling character, and reveals its contents near the very end of the story to a repulsed and amazed Kipling: "He fumbled in the mass of rags round his bent waist;… and shook therefrom on to my table—the dried, withered head of Daniel Dravot!"

So he'd been carrying this withered head with him on his journey from Kafiristan. It's a distracting and ultimately unhelpful climax in a sense, because it displaces our attention from the story's critique of imperialism to this horror-film shocker. Even worse, the severed head and the fact that Peachey is nearly killed (Peachey describes how he is literally crucified—the natives crucify him in punishment, and then when he doesn't die they cut him down from the tree on which he's been crucified and send him on his way), so the severed head and Peachey's crucifixion have the awkward effect of seeming to valorize a view of the natives as more barbaric and savage than their conquerors.

So the story loses focus in the end, partly undermining or contradicting its primary fable. But damaged or disappointing endings are common even in great stories; the *Odyssey* is an example. However it falters in its final pages, *The Man Who Would Be King* remains a landmark of English literature, as disturbing and instructive today as it was nearly 120 years ago, when it was first published.

Lecture Six
Heart of Darkness—Europe's Kurtz

Scope:

This lecture begins with a sketch of Conrad's life, emphasizing the ways in which his orphaned childhood and life-long exile from his native Poland shaped his fiction. Conrad's discovery of his first-person narrator Marlow released his gift for creating vividly concrete moments that were simultaneously symbolic and deeply allusive. This mingling of the particular and the general or universal is a key as well to the character and meaning of Kurtz in "Heart of Darkness," a "universal genius," as some call him, who embodies the missionary and corporate ambitions of Western culture. "All Europe went into the making of Kurtz."

Outline

I. We begin with a brief look at Conrad's life.
   A. Joseph Conrad (1857–1924) was born in the Polish Ukraine, then part of Czarist Russia.
   B. His mother died when Conrad was eight. His father, Apollo Korzeniowski, died when Joseph was 12, and the young Conrad led a politically forbidden funeral cortege down the streets of Krakow, as thousands of Polish patriots mourned Apollo's death.
      1. Not only a Polish nationalist, Apollo was a fervent Christian who wrote poetry that imagined Poland as the crucified Christ.
      2. Conrad left Poland at 16 and spent the rest of his life in exile. He was haunted all his life by a sense of having betrayed his country, and the themes of betrayal and loyalty, as well as distant or absent fathers and surrogate elders, are central to his novels.
   C. Conrad spoke Polish, French, and English with a thick Polish accent and was a sailor in the 1880s.
   D. In his biography *Joseph Conrad: The Three Lives*, Frederick Karl writes of Conrad's three separate identities, which

never easily coexisted: Conrad as a Pole, a sailor, and an English writer.
- **E.** Conrad was already middle-aged when he started writing, and he consciously imitated Gustave Flaubert, who was a great influence. His early pieces are interesting but only as apprentice work; they seem alien from his later fiction.

II. The most important event in Conrad's life was the discovery of his narrator Marlow, who first appears in a few stories in the late 1890s. One of these stories is *Heart of Darkness*.
- **A.** Marlow, an Englishman, first appears in Conrad's short story "Youth," about a young seaman who is forced to take command of a lifeboat that has been cast loose from a sinking ship.
- **B.** Following "Youth," Conrad began a short story that grew to become the novel *Lord Jim* (1900). Marlow is also the narrator in this book; and while Conrad worked on the story, he also began *Heart of Darkness*, which he completed before *Lord Jim*.
- **C.** As *Heart of Darkness* begins, Marlow sits on the deck of a yacht moored on the Thames River.
    - **1.** He thinks of empire and appears to distinguish between British and Belgian imperialism, which is understood to be far more monstrous than the English variety.
    - **2.** Marlow implies that certain forms of imperial rule are morally superior to others. Whether or not Conrad believed this himself is unclear, but Marlow seems to have believed it.
- **D.** Marlow's voice liberated Conrad because it enabled him to distance himself from the narrator and from the material of the story.
    - **1.** Psychological critics have suggested that this technique was a necessity for a man who was alienated at least twice from his homeland.

III. Marlow seems to have released Conrad's distinctive tone, especially his capacity to represent moments that are simultaneously vivid and particular *and* symbolic or mythic.

    A. The knitting women at the beginning of the story, for example, are meant to suggest the knitting Fates, reminding us of the pervasiveness of mythic elements in Modern literature and illustrating Conrad's ability to bring together the particular to the general.

    B. As we saw in "Captain Carpenter" in Lecture One, some Modernist texts invoke the past in a spirit of parodic satire, in which the comparison with the past judges or measures the disorder of the present.

    C. In *Heart of Darkness*, the mythic references and undertones complicate and universalize Marlow's individual journey, which is seen to resemble or reenact the epic descents into the underworld of such heroes as Odysseus, Aeneas, and Dante.

    D. Marlow's trip up the Congo is a descent in a political, moral, and psychological sense, and it reveals man's capacity for savagery and evil.

    E. The references to the knitting Fates illuminate Marlow's journey but in an ironic way, for his journey involves no rebirth or redemption. The mundane physical reality of the knitting ladies—with warts on their faces and cats in their laps—undercuts the scene in a way that is characteristic of Conrad's ironic sense of life.

    F. The way in which the real and symbolic coexist in Conrad's work, even within the same passage, is one of the keys that makes his writing memorable. In the preface to one of his novels, he articulates his mission: "My task above all is to make you see."

IV. The other main character, Kurtz, is also important both literally and symbolically.

    A. Kurtz's physical appearance has both a vivid concreteness and a terrible symbolic implication. Although Marlow imagines him throughout the story as a powerful character, when the two finally meet toward the end, Kurtz is a mere specter, suffering from disease and appearing ghostly, as if he belonged in a nightmare.

    B. Kurtz thus embodies the autobiographical and psychological dimensions of *Heart of Darkness*. Conrad, too, had

journeyed up the Congo River and said that the events in the story were merely extrapolations of his own experiences.
C. One reason Marlow is reluctant to reach Kurtz is that Kurtz represents impulses to savagery that Marlow recognizes, to some degree, in himself.
1. For Marlow, Kurtz is a symbolic double.
2. Marlow is charged with bringing Kurtz back to civilization, but Kurtz resists. At one point, the men wrestle, an act which suggests that Marlow literally and figuratively embraces his double.
D. Early in the story, Marlow reminds us of the extent to which he identifies with Kurtz. The latter's mission in the Congo is a moral quest to "civilize" the "savages," a quest which Marlow admires.
E. We come to realize that all Europe contributed to the making of Kurtz in a deep and terrible way. Kurtz is an emissary of Europe's best possibilities, and his life and last words are a profound commentary on the imperial adventure.

Permissions Acknowledgment:

"To the Reader" by J.V. Cunningham from *The Poems of J.V. Cunningham,* Steele, ed. Copyright © 1997. Reprinted by permission of Swallow Press/Ohio University Press, Athens, Ohio (www.ohioswallow.com).

Essential Reading:

Conrad, *Heart of Darkness.*

Supplementary Reading:

Conrad, *Lord Jim.*
Guerard, "The Journey Within," in *Conrad the Novelist,* pp. 35–48.

Questions to Consider:

1. Why does Conrad evoke earlier stories by Homer, Virgil, and Dante that deal with descents into the underworld? How does Conrad differentiate Marlow's journey from those of his Classical ancestors?

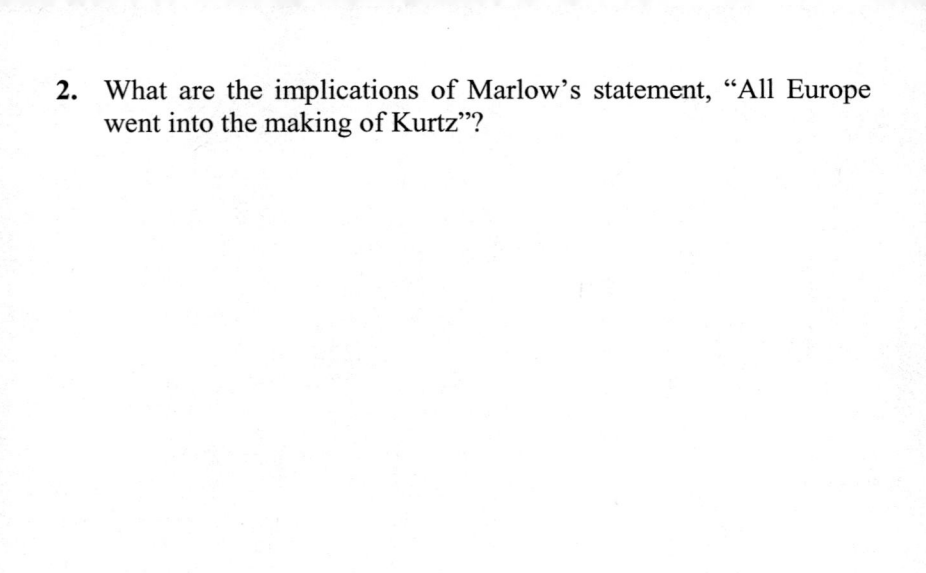

2. What are the implications of Marlow's statement, "All Europe went into the making of Kurtz"?

Lecture Six—Transcript
Heart of Darkness—Europe's Kurtz

Joseph Conrad, one of the truly unique figures in the history of Modern literature, was born in 1857, before our Civil War and before the emancipation of the serfs in Russia, and died in 1924. In addition to being a unique writer, a writer whose works have a unique feel, a unique tone, his life itself had unusual qualities, to put it mildly.

He was a Pole. He was born in Poland; spent his childhood in Poland. His father was a political activist and Polish nationalist. Poland during Conrad's childhood was under the control of Russia, and his father was exiled along with his family—with his mother and the young baby Joseph—to Siberia when Conrad was a young child. His mother actually died in exile in Siberia when Conrad was eight years old. And his father was then allowed to return to Poland, and died when Conrad was twelve. So he was something of an orphan even for a good part of his childhood, and lived for part of that time in exile in Siberia.

One of his most powerful early memories was a memory of having walked in the front, leading a politically forbidden funeral cortege down the streets of Krakow, as thousands of Polish patriots mourned the death of his father Apollo, and implicitly in their mourning protested Russian rule. His father, as I said, was a Polish nationalist and also a fervid Christian who wrote poems and stories about Poland as a form of the crucified Christ.

Conrad was haunted all his life by a sense of having betrayed his country, and the themes of betrayal and loyalty and distanced or absent fathers and surrogate elders are central themes in his novels.

His second language wasn't even English—although he became a great English novelist—but was French. Most educated Poles of his aristocratic class learned French as children, and Conrad was no exception. He learned English, his third language, only as an adult when he was living in France, when he got work as a sailor and began sailing on English vessels in the 1880s. He spoke with a thick Polish accent all his life, and the accent was said to have grown thicker with age.

One of his biographers, Frederick Karl, titled his book *Joseph Conrad: The Three Lives*, his implication being that there were three separate identities, which never existed or coexisted very easily within Conrad's personality. He was a Pole and son of Apollo Korzeniowski, the Polish nationalist. He was the sailor who sailed on a number of vessels and was the captain of a sailing vessel for a brief time in his career as a seaman; and then he became the English writer. And Karl's theory, I think a powerful one, is that those three different dimensions or sides of Conrad's personality never came into full congruence, and that some of the jagged ambiguity and haunted quality of so many of Conrad's great novels are partly a function of that divided self.

The most critical and important event in Conrad's literary life was the discovery of his narrator Marlow. He began his career—already in middle age when he turned to writing—in imitation of certain 19th century originals, and in particular in imitation, although he wouldn't have called it conscious imitation, of the great French novelist Flaubert, who was a great influence on many English writers.

Conrad read Flaubert as a young man and was very familiar with his work, and Conrad's earliest stories and novels use a kind of Flaubertian irony, and a kind of omniscient narrator who is distanced from and in many ways very unfriendly toward his characters. Those early works of Conrad's (only a few of them) are interesting, but only as apprentice work; and they're especially interesting to students of Conrad today because they seem so alien in some sense, so different from his great fiction.

The great discovery that Conrad made relatively early in his writing career, sometime in the 1890s, was the discovery of this first-person speaker, this Englishman named Marlow who narrated the story to others. Marlow first appears in a short story entitled "Youth," a story about a young seaman who is forced to be the captain of a lifeboat, of an escape boat from a ship that's sinking. And while he was working on "Youth," certain other stories began to occur to Conrad, and he then began a second short story that kept getting longer and longer, and ultimately became the great novel *Lord Jim*, published right after the turn of the 20th century.

And while he was working on *Lord Jim*, which kept getting out of control and getting larger and larger, he stopped again and began his work on *Heart of Darkness*, which he actually ended up finishing first. Both "Youth" and *Heart of Darkness* were published in a magazine in Scotland called *Blackwood's Magazine*. It was a magazine that often published exotic local-color stories. It had published Robert Louis Stevenson, for example, and it published stories that were in imitation of local-color stories by writers like Kipling.

And it's often said by Conrad scholars that the readers of *Blackwood's Magazine* would not have been able to recognize the deeply ironic and anti-imperial tendencies in stories like *Heart of Darkness* and *Lord Jim*, because they appeared in a magazine environment in which such irony was never available, was never seen.

This may explain a certain confusing passage at the beginning of *Heart of Darkness* where Marlow, sitting on the deck of a cruising yawl (of a yacht) in the mouth of the Thames, is talking to his listeners and begins to talk about the different places in Africa that are ruled by different colonial powers, and he seems to be praising the British. He says, ah, wherever we see red on the map—which is the color of the British imperial colonies—we know "some real work is [being] done … there"; but I didn't go into the red in the story I'm about to tell you. I went into the terrible yellow—which was the Belgium Congo.

Marlow seems in that opening paragraph to be distinguishing British from Belgian imperialism, to be implying that certain forms of imperial rule are morally superior to others. And it may well be that Conrad himself believed this, although it seems to me more sensible to think that his character Marlow may have believed this, but that the author of *Heart of Darkness* could not possibly have thought such a thing, so thoroughgoing and systematic is *Heart of Darkness's* attack on the very foundations of imperial pretensions, of the imperial vision.

In any case, the discovery of Marlow was an immensely liberating one for Conrad. One of the reasons for this seems to be that he needed the screening or distancing effect that a first-person narrator, not himself, would allow. It's almost as if it seems that Conrad had difficulty describing action directly, but that if he could describe it as something that had happened in the past that was being recalled by another character, he was able to encompass the material, to confront the material in a more dramatic way. And there have been psychiatric critics, psychological critics who have suggested that this was a necessity for a man who was at least twice alienated from his homeland.

Here's an example of Marlow's special voice, and I hope it will be an example of how Marlow's voice liberated Conrad into a form of discourse that allowed him simultaneously to be profoundly concrete and vivid, but at the same time deeply symbolic and generalizing. It's one of the deep secrets of Conrad's writing, I think, and one can hear it even in simple passages. Here's a passage from relatively early in the story, when the young Marlow goes to the company offices in Belgium to receive his commission to travel up the Congo River on his imperial mission.

He enters the outer office and he encounters two women who are knitting in the outer office, and it becomes clear as you listen to the passage that these knitting women have a kind of symbolic power, a symbolic force. We come to realize as we're reading the passage that they stand for the knitting Fates of classical mythology. I'm going to read from the second reference to them. He arrives. He meets the knitting women, who welcome him. He's sent briefly into an inner office where he stays for a very brief time and then he comes back out; and when he re-emerges, he sees these knitting Fates again.

> In the outer room the two women knitted black wool feverishly. People were arriving, and the younger one was walking back and forth introducing them. The old one sat in her chair. Her flat cloth slippers were propped up on a foot-warmer, and a cat reposed on her lap. She wore a starched white affair on her head and had a wart on one cheek and silver-rimmed spectacles hung on the tip of her nose.

What could be more concrete, more specific, more homey, in some ways almost comic? But then watch how the paragraph opens out.

> She glanced at me above the glasses. The swift and indifferent placidity of that look troubled me. Two youths with foolish and cheery countenances were being piloted over, and she threw at them the same quick glance of unconcerned wisdom. She seemed to know all about them and about me too. An eerie feeling came over me. She seemed uncanny and fateful. Often far away there I thought of these two, guarding the door of Darkness, knitting black wool as for a warm pall, one introducing, introducing

continuously to the unknown, the other scrutinizing the cheery and foolish faces with unconcerned old eyes. Ave! Old knitter of black wool. *Morituri te salutant.* [The line that the gladiators in ancient Rome would say: "We who are about to die salute you."] Not many of those she looked at ever saw her again—not half, by a long way.

Well I hope you can feel what a remarkable passage that in fact is. It suggests something to us about the way myth in general works in Modern literature, as well as Conrad's habit of moving from the particular to the general with a great swift clarity and power.

These knitting Fates that welcome young Marlow to the offices of the company that will send him to the Congo and to the heart of his own and his culture's darkness remind us of the pervasiveness of myth in Modern literature.

We saw in "Captain Carpenter" that some Modern texts invoke the past in a spirit of parodic satire, in which the comparison with the past judges or measures the meanness, the disorder of the present.

This is how myth works in T.S. Eliot, in poems like "The Wasteland" or "Sweeney Among the Nightingales." And it is how myth works in some degree in Conrad. But I want to mention parenthetically something we will confront much more fully later on in these lectures: This is *not* how myth works in James Joyce, *not* how myth operates in *Ulysses*, which has an even more complicated attitude toward its relation to its mythic ancestry.

In *Heart of Darkness* the mythic references and undertones, which are not restricted only to this one moment in the text at all, complicate, deepen, universalize Marlow's individual journey. What we're reminded of in this passage and then in a number of other passages that make some of these associations explicit, is that there is an epic convention of the descent into hell in which the epic hero, Odysseus or Aeneas, or in the Middle Ages, the character of Dante in the *Divine Comedy*, make a descent into the nether regions, into hell; and when they emerge from that descent they're changed in some fundamental way.

And it becomes clear that those mythic parallels are being re-enacted in some sense, that Marlow's trip up the Congo River is a descent in a way into the darkest possibilities of human nature in both a political and a moral sense, and that it also becomes, as I'll try to indicate in a moment, a psychological descent into one's own self, into one's own capacity for savagery and evil.

The references to those knitting Fates have that kind of quality about them even in the beginning. They suggest, they enlarge or mythologize Marlow's journey, but they do so in a very ironic and partly compromised way, because Marlow's descent or quest is an ironic one. It eventuates neither in rebirth nor redemption. And these knitting Fates have warts on their noses and cats in their laps. There's a kind of mundane undercutting in the scene that is characteristic of Conrad's way of dealing with these mythic resonances.

The way in which the real and the symbolic coexist simultaneously, within the same passage often in Conrad's work, is one of the keys to what makes his writing so resonant and so memorable. Even the vivid physical descriptions in *Heart of Darkness* have this quality. It's easy to overlook this side of the novella even though it's a tremendously important one—what we might call the travelogue aspect of *Heart of Darkness*.

In the preface to one of his novels Conrad articulated a kind of mission statement, and one of the most famous lines, it's actually one of the most famous lines in all of Modern literature, is this one that Conrad makes in that preface. He says, "My task, above all, is to make you see." My task is to make you see, above all. And he had a deep commitment to trying to create vivid scenic effects that would be deeply memorable. No reader of *Heart of Darkness* can forget some of those moments, but what's especially interesting about so many of those moments is that they are simultaneously profoundly realistic, grounded in vivid details that make them visually memorable, but also carry symbolic overtones, as if the symbolic and the realistic are organically linked in Conrad's writing.

Let me give you one example of this from among many, many examples. This is a passage, a relatively minor passage in a way, from a moment in the text where Marlow is in the heart of darkness and he notices an object. He says:

> I came upon a boiler wallowing in the grass, then found a path leading up to the hill. It turned aside for the boulders, and also for an undersized railway-truck lying there on its

back with its wheels in the air. One was off. The thing looked as dead as the carcass of some animal. I came upon some more pieces of decaying machinery, a stack of rusty nails. To the left a clump of trees made a shady spot, where dark things seemed to stir feebly. I blinked, the path was steep. A horn tooted to the right, and I saw the black people run. A heavy and dull detonation shook the ground, a puff of smoke came out of the cliff, and that was all. No change appeared on the face of the rock. They were building a railway. The cliff was not in the way or anything; but this objectless blasting was all the work going on.

Now that reference in that passage to the motions that he can't quite recognize is actually a reference to a passage even more famous than the one I've just read to you that comes a few paragraphs later that is known in the literature as the "grove of death" scene, and of course this is a passage in which Conrad's hero Marlow comes across a group of native workers who are now so weary and so starved that they can no longer do any more valuable work, and they have in a sense gathered in this grove to die.

These passages and many others like them in *Heart of Darkness* have a tremendous vivid concreteness, an immediacy that I associate with what we might call the travelogue dimension of *Heart of Darkness*, the way of introducing readers to what the physical experience of being in Africa was like. And there's no question that this was one of primary impacts that the novella first had when it was first published.

But in that passage that I just read to you about the boiler wallowing in the grass and the railway car lying upside down on its back, we also have a profound deconstructionist image or series of images of the uselessness, of the folly of imperialism; because here are the implements, the machinery of Western civilization, being engulfed by the darkness, being engulfed by the jungle, being turned useless in a certain sense by the jungle. And the objectless blasting that Marlow describes is in part, I think, to be associated with the objectlessness or purposelessness of the futility of the imperial adventure, of what he calls at one point the "rapacious and pitiless folly" of the conquest of Africa.

There are numberless passages—maybe not quite numberless; but many, many passages—in *Heart of Darkness* that have an equal vividness and also an equal symbolic significance, an equal symbolic charge. The boiler in the grass is an example. The grove of death is an example. Those of you who've read the story recently may recall another very eerie passage. Before Marlow even arrives at the heart of darkness he sees a French gunboat shelling the continent of Africa, and he describes the gunboat firing these shells: Poof, go the guns, Marlow says, and the shells fly through the air and land with a puff of smoke someplace in the interior, but all that happens is the puff of smoke; there's no other response. It's a version of the objectless of blasting. It seems purposeless. It seems useless.

So this vivid physical reality and the symbolic suggestiveness go together. And one of the things that's implied by the images I've already described is something that becomes explicit in Marlow's language at a number of points, which is not just that this is eerie and terrible and purposeless, but it also is almost as if it's a kind of dream or a nightmare, reinforcing the idea that we're making a kind of descent into hell. Even more than that—not just a kind of descent into hell—but that this suggestion of nightmare implies something that the story makes explicit at various points.

Marlow says at one point, "We live, as we dream—alone." At another point he says it seemed as if I was doing this as if in a dream. And the idea that what's happening here has a kind of dreamlike quality takes the story into another dimension as well. I can get at some of that dimension by talking about the character of Kurtz, the central symbolic figure in the story, and I want to spend the rest of this lecture saying a few things about that complicated figure.

One of the things we can say about Kurtz is that he operates in the story something in the same way as these physical details I've mentioned, because Kurtz's physical appearance has both a vivid concreteness and a terrible symbolic implication. For example, he's described, when Marlow finally gets to see him—in his narrative he keeps postponing our arrival to Kurtz, and in addition to that postponement we also keep hearing about Kurtz from other characters. There are even some readers who are kind of disappointed when we finally reach him because there's been such a buildup. So many characters in the story have ideas and attitudes about him.

And Marlow himself, at one point in the story when he hears a rumor that Kurtz might be dead and he says, 'I was devastated,' (I'm paraphrasing; he doesn't use the word "devastated" but he means it by implication). 'I was very upset by the idea that Kurtz might be dead. I realized that I had been incredibly looking forward to meeting him. I needed to see him. He was the object of my quest in some respect.'

When he finally does meet Kurtz he finds that he's almost a wraith. He's described as a wraith, as a ghost, as someone spectral, and he has a bald head; all his hair has fallen out. He's in the final phases of malaria or some other terrible disease that he's picked up in the jungle, and he's emaciated and gaunt and wraithlike and ghostlike, as if he belongs in a nightmare or a dream rather than in reality. And his bald head also suggests something else about him, because one of the things he is, of course, is an ivory hunter, and it's as if he's become the thing he has sought, as if he's become the object of his quest.

So, there is a kind of vivid evocation of him, but the very same details that evoke him also suggest symbolic implications, that he's a figure of dream or nightmare. And this idea that Kurtz may be a figure of dream or nightmare leads us more closely into what might be called the psychological or even in one way the autobiographical dimensions of *Heart of Darkness*. Conrad himself did make a journey up the Congo River something like what he describes in *Heart of Darkness*. And he mentions in one of his prefaces that the story of *Heart of Darkness* pushes his autobiography only a little way, implying that his real experiences were not that different from the story as it is described.

In the trip up the Congo River toward Kurtz we constantly hear words about who Kurtz is and what Kurtz looks like, and we begin to realize, I think, it becomes explicit in some passages, that Marlow regards Kurtz as a kind of symbolic double, as a possible self. And one of the reasons that Marlow was reluctant to reach Kurtz, and one of the reasons that his narrative almost seems to be evasive—when it comes to the point of almost describing him, Marlow backs away and comes up with some kind of digression to keep himself from reaching this terrible location—one suggestion that I think is widely accepted now in the scholarship is that one of the reasons Marlow was so reluctant was that Kurtz represents certain possibilities of savagery and evil that Marlow recognizes in some degree within his own breast. The classic account, the first defining account of this symbolic aspect of *Heart of Darkness* as a story about a descent into oneself, as a story about a confrontation with one's own nature, was made by the great Stanford critic Albert J. Guerard, who wrote a famous essay called "The Journey Within" in which he systematically went through *Heart of Darkness* pulling out all the references that I've alluded to here to show how closely and systematically the story mapped a kind of account of a descent within one's own body, within one's own soul.

And this symbolic journey reaches its climax in some sense in the moment when Marlow finally encounters Kurtz, and Kurtz is riven by illness and doesn't want to be taken back to civilization, which is Marlow's mission. He has to get Kurtz and bring him home. And Kurtz crawls away from him in the middle of the night, sick as he is; he can't walk, he's too ill. And he crawls out into the jungle and Marlow has to follow him into the jungle, and in this climactic moment Marlow actually grabs Kurtz and they have a kind of wrestling match.

They're rolling around on the jungle floor, and Marlow says, "[I had] to deal with this shadow by myself alone." And while he's doing this wrestling match, he confounds the beating of his own heart with the drums that he hears in the jungle—from the natives who are beating their drums. Kurtz has become a god in the land, and the natives don't want to allow Kurtz to leave.

Well, that Marlow's own heartbeat is confused with the sound of the jungle suggests that Marlow, at this moment (in addition, the wrestling also does this) he's literally embracing his double in a certain way—suggests the danger and the fascination of Kurtz for Marlow. So, that wrestling match in the jungle suggests again this doubling.

So, there is a deep personal and psychological dimension to *Heart of Darkness* and to Kurtz's meaning. But there's another aspect to Kurtz, maybe even more important in some way, and we can get at that aspect by recognizing again the extent to which Conrad's language gives us both concrete and symbolic details simultaneously.

And I have another example of this from another passage that I would like to read to you. There's a powerful moment in the story in which Kurtz is described in apparently a very gentle way. It says: "The original Kurtz had been educated partly in England, and—as he was good enough to say himself—his sympathies were in the right place. His mother was half-English; his father was half-French. All Europe contributed to the making of Kurtz." "All Europe contributed to the making of Kurtz," and as the story goes on, we come to realize that all Europe contributed to the making of Kurtz in a deep and terrible way, that Kurtz becomes a kind of emissary of Europe's best possibilities.

Marlow very early on in the story reminds us of the extent to which he identifies with Kurtz against all the other "pilgrims" who work for the company. He uses the term pilgrim in an ironic sense, mocking them because he knows they're the opposite of pilgrims. They're just out there for loot. But Kurtz is special because he went out there, Marlow tells us, "[armed] with moral ideas of some [kind]," armed with a larger mission. Among other things, Kurtz is a representative of the International Society for the Suppression of Savage Customs. That is to say, he has taken onto himself the white man's burden of transforming this benighted savage area into a civilized one.

And in his report to the Society for the Suppression of Savage Customs, which Marlow comes across when he goes through Kurtz's papers, he finds the pages that Kurtz has written "vibrating with eloquence," he says, and then at the very bottom Kurtz has scribbled in a desperate scrawl, "Exterminate all the brutes!" That detail itself suggests something of Kurtz's own decline, of Kurtz's own descent into savagery. And, indeed, all Europe, from that angle, went into the making of Kurtz.

There are many other details in the story that emphasize Kurtz's remarkable powers. He's associated with painting. He's a painter; he's a poet. At one point late in the story Marlow discovers that some people thought that he might have been a politician "on the popular side." And, of course, he's a great explorer and the greatest of all ivory carriers. It turns out that we come to recognize that Kurtz represents the best of European civilization, what's most remarkable about European civilization, and that what is most remarkable about European civilization—it's highest achievements—should descend into savagery, should have become such a murderer, should have become such an instance of "the horror" (Kurtz's last words), is a deep commentary on the imperial adventure.

And we'll continue with these themes and link them to Marlow's very complex and self-conscious narrative strategies in my next lecture.

Lecture Seven
Heart of Darkness—The Drama of the Telling

Scope:

The most distinctive and, in some respects, the most frustrating aspect of *Heart of Darkness* is the story's strange and apparently digressive structure. This lecture centers on the novella's dislocated chronology and self-conscious first-person narrator. Marlowe's tentative, back-tracking, interrupted narrative creates two separate "stories"—one story is the "traditional" adventure or action, the second a running commentary on that action and the difficulties of retrieving it in memory and in words: a *drama of the telling*. This characteristic Conradian drama of self-division and self-consciousness has counterparts in nearly all the modernist works that lie ahead of us in this course. It is a literary equivalent of the self-reflexive energies of the Impressionist and Postimpressionist painters, a distinctive theoretical dimension of nearly all the signal achievements of high Modernism. The lecture concludes by speculating on the moral and political ironies in Conrad's ending, where the dead Kurtz's fiancée coerces Marlow to disclose her idol's dying words.

Outline

I. Conrad's most distinctive contribution to Modernist literature is his narrative strategy. This structure in *Heart of Darkness* is challenging to the reader.
   A. Powerful similarities exist between Marlow and the narrator of Samuel Taylor Coleridge's poem "The Rime of the Ancient Mariner."
      1. Like the Ancient Mariner, Marlow is obsessed yet not in control of the direction or connections that he makes through the course of the story.
      2. Marlow has a memorable eloquence, but there is also a pathological quality to his insistent postponement of a confrontation with Kurtz.
   B. The structure of *Heart of Darkness* is a series of approaches and withdrawals in which Marlow promises to take us up the

river to reach his disturbed double, Kurtz, but backs away as he gets close to the task.
- C. This narrative structure displaces our attention from the traditional narrative matter of the story and toward the drama of the telling of the story. Conrad foregrounds the act of narration itself: one of his central themes is the problem of telling the story.
- D. As Marlow continues his fragmented and digressive tale, we realize that he is preoccupied with the difficulty of recapturing what happened. In certain moments in the story, another framing narrator takes over when Marlow pauses.
- E. Marlow seems either unable to find conclusive meaning in his experience or reluctant to acknowledge its meaning.
- F. There is a deep sense that Marlow needs to tell the story again and that his material is so psychologically and morally disturbing that the tale will always be unfinished.
- G. Thus, *Heart of Darkness* calls attention to certain aspects of the difficulties of writing. We might think of this as the literary equivalent of the self-consciousness in certain forms of Impressionist and Postimpressionist paintings by Van Gogh and Monet (discussed in Lectures Three and Four).
- H. In *Heart of Darkness* and Conrad's other works, the narrator confesses his failure and is at times so frustrated that he even attacks his listeners.
- I. Marlow's fear that he is like Kurtz haunts his narrative and is one of the reasons that the narrative is evasive and temporally fragmented.

II. Marlow's personal anxieties and needs shape the narrative, but so do other elements.
- A. The story takes on an epistemological or philosophical dimension, dramatizing the problem of knowledge, of the limitations of our cognitive powers.
- B. This approach also dramatizes the limits of language as an instrument for representing experience.

III. A summary of the novel's ending will recap its major themes.

A. Toward the end, Kurtz lies dying on the steamboat as Marlow attempts to bring him back to civilization. Kurtz's final words are "The horror! The horror!"

B. When Marlow returns to England, he meets various acquaintances of Kurtz, many of whom glorify the deceased adventurer's character and what he might have aspired to had he lived.

C. Marlow also meets Kurtz's fiancée, who believes in Kurtz's nobility and is devoted to the mythology of Kurtz as the great Christian do-gooder who has gone into the heart of darkness to transform a savage civilization. She is committed to the idea that Kurtz's memory must be preserved because of his remarkable nobility.

D. The Kurtz she imagines does not exist at all. The real man is virtually the opposite, taking a native mistress, and killing others at will. He is a man who (paraphrasing Marlow) had kicked himself loose from the Earth, a man who became a god in the land, an embodiment of appetite and savage unrestraint.

E. When Marlow encounters Kurtz's fiancée, he knows all these terrible things about Kurtz. He especially remembers Kurtz's last words. Marlow is horrified by the fiancée's deluded notions, but she coerces Marlow into reassuring her of Kurtz's nobility.

F. Such coercion is a recurring drama in Conrad, and it shows his insight into the idea that conversation itself is political and coercive, not necessarily a mutual experience.

G. The fiancée asks Marlow Kurtz's final words. Marlow yearns to answer truthfully. His whole narration has been about truth and the difficulty of reaching it.

H. But as much as Marlow hates lies (as he tells readers earlier), he tells the fiancée what she wants to hear—that Kurtz's last words were her name.

I. In an ironic twist, however, the fiancée believes wholeheartedly in Kurtz's nobility and thus completely believes and embraces the rationale for Western imperialism.

J. The ending implies, in a wonderful and disturbing way, that Marlow's lie speaks a kind of truth. No wonder the fiancée is confined, in the story, to a sepulcher and is, for the reader, a kind of horror herself.

Supplementary Reading:
Conrad, *Nostromo*.

Questions to Consider:
1. Why is Marlow especially preoccupied by Kurtz's eloquence and by the fact that he "presented himself to me as a voice"?
2. What is the effect of Marlow's reluctance or inability to tell his story in chronological order and his tactic of interrupting himself and returning us to the scene of his narration?

Lecture Seven—Transcript
Heart of Darkness—The Drama of the Telling

Dr. Samuel Johnson, the famous dictionary maker and 18th-century philosopher, famously said, "If one read Samuel Richardson for plot, one would hang oneself." And there have been similar responses to the narrative structure of Conrad's *Heart of Darkness*. If one reads *Heart of Darkness* with the expectation that one is getting a conventional kind of adventure story or local-color story, one is bound to be disappointed, because the most distinctive and in some respects for some readers the most frustrating aspect of the story is the strange and digressive narrative strategy on which it's based.

I want to devote this lecture primarily to that strategy, which is one of Conrad's most distinctive contributions to Modernist literature, and has the profoundest of ramifications for Conrad's art.

We might begin, however, by thinking of a kind of comparison between Marlow the narrator and a much earlier mariner. Can you guess who is being referred to here?

> Forthwith this frame of mine was wrenched
> With a woeful agony,
> Which forced me to begin my tale;
> And then it left me free.
>
> Since then, at an uncertain hour,
> That agony returns:
> And till my ghastly tale is told,
> This heart within me burns.
>
> I pass, like night, from land to land;
> I have strange power of speech;
> That moment that his face I see,
> I know the man that must hear me:
> To him my tale I teach.

Well, of course, that's Coleridge's ancient mariner, from relatively early in the famous poem "The Rime of the Ancient Mariner." And there are powerful similarities between Marlow and the ancient mariner. Marlow's maybe not quite so ancient as the mariner, but like Coleridge's character, he is obsessed, driven. One feels as one reads his narrative that he's not totally in control of the direction or connections that he makes in the course of the story. He does seem to have a "strange power of speech," a remarkable and memorable eloquence, but there's also a sense that there's an obsessional and even perhaps a pathological quality to his insistent postponement of his confrontation with Kurtz, to his apparently almost endless series of strategic reversals or denials or backings away. One might describe the structure of the story as a series of approaches and withdrawals in which Marlow promises to take us up the river to reach his disturbed double, Kurtz, but the minute he gets close to the task of actually describing him, he backs away and finds some excuse to digress into some other topic.

After a while this habit of the narrative becomes, in a certain sense, a structural principle, and one comes to realize that something very significant is going on. What this narrative structure does, among other things, is displace the reader's attention from what might be called the traditional narrative matter of the story toward something else, something I call the drama of the telling of the story. It's as if what Conrad has moved into is a form of storytelling which foregrounds the act of narration itself, which takes as a central subject matter (not as a minor aspect of the story or as an afterthought or as a minor implication, but as a central element, as a central fundamental part of the drama) the problems of telling the story.

Once one recognizes that Conrad displaces the traditional narrative matter of the story, we can come to understand more clearly what the story is actually doing, why this drama of the telling is at the center of Conrad's vision of life and at the center of the impact that *Heart of Darkness* makes upon its readers.

As Marlow continues his fragmented and digressive and almost repetitive (he doesn't actually literally repeat himself, but he certainly approaches certain topics again and again, so there's an appearance, at least, of repetition), as Marlow does that, we begin to realize a couple of things. First, that Marlow is preoccupied, literally, by the difficulty of talking, by the difficulty of recapturing what happened. There are certain moments in the story where he interrupts himself and he stops talking, and we are brought back to the scene of narration.

Remember, the novella opens with Marlow sitting on the deck of a cruising yacht in the mouth of the Thames, and the first narrator is a framing narrator—the same principle, in a way, as "The Man Who

Would Be King." The framing narrator talks about Marlow, and talks about Marlow being famous for inconclusive yarns, stories that seem to not have clear meanings; and of course there is a kind of penumbra of symbolic implication that surrounds nearly all the details of Marlow's narration, but he does seem to be a fellow who is narrating something whose conclusive meaning he himself is reluctant to acknowledge or reluctant to find.

And there's even a deep sense as Marlow continues his narration that he's not finished, that he needs to tell the story again, that his narration is in that deep sense inconclusive, unfinished, and will perhaps always be unfinished, so fraught, so dangerous, so psychologically and morally disturbing is the subject matter with which he's dealing.

So, what I'm suggesting is that one of the really distinctive inventions of *Heart of Darkness* is this drama of the telling of the story, which not only foregrounds the act of telling, but which in a systematic way calls attention to certain aspects of the difficulties of writing. One might think of this as the verbal equivalent, the literary equivalent, of the self-consciousness I called attention to when we were talking about certain forms of Impressionist and Postimpressionist painting. When van Gogh calls attention to the act of painting by creating thick stabs of paint on his canvas put there by his thumb, or when Monet calls our attention to the way in which that façade of the cathedral will change under different lights, and we have the sense that the appearance and significance of the cathedral is constantly in danger, is constantly threatened, is constantly changing, that's a kind of painterly or visual equivalent of the sort of thing I'm saying is true of *Heart of Darkness*.

And of course it's not only true of *Heart of Darkness*. Conrad pursues similar strategies in his great novels—in *Lord Jim,* in *Under Western Eyes*, in the novel *Chance*—which are also novels that are narrated by particularized narrators. In the case of *Lord Jim* and *Chance*, Marlow returns to do further duty as a Conradian narrator. In *Under Western Eyes* it's a series of other narrators, including the primary narrator who's a Western teacher of languages.

And in all these texts one of the primary functions of the narrator is to confess his failure, is to say "words fail me." And if you look closely at *Heart of Darkness* you will find that there are many moments in the narrative where Marlow interrupts himself and stops talking, and the framing narrator—the one I mentioned who starts off the story and then yields to Marlow and to Marlow's narration—sometimes then will come back in again and say, Marlow ceased and smoked his cigar for a moment and in the darkness all I could see was the glowing end of his cigar.

At certain times when Marlow interrupts himself he actually attacks his listeners—not a very good strategy for a narrator who wants to be listened to and respected—but the implication is that Marlow is so frustrated in his desire to communicate what he experienced, what he knew in the heart of darkness, that he becomes angry at his listeners. At one point he says, How can you understand me? You who are here in civilization performing your "monkey tricks"—the implication being that the ordinary life of those of us who live in safe and civilized places are merely trivial, are merely a series of monkey tricks; that the real danger, the real darkness, the real savagery that lurks beneath the surface of all human possibility is simply held in check, is simply hidden by the façade of clothing and courtesy and procedure that organizes civilized life.

And that what happens to Kurtz when he goes into the heart of darkness, what nearly happens to Marlow when he follows him there, is the falling away of these constraints of civilization, leaving Western man, European man, in confrontation or in conversation with his most primitive, with his most fundamental, with his untrammeled self. And Marlow fears that Kurtz's example—this dangerous doppelganger, this double of his—represents not only a unique possibility, but the possibility that is resident within his own heart, and by implication resident within the psyches, within the souls, within the hearts of all people who live in civilized societies; as if Kurtz, who represented the highest possibilities of civilization, also in a most fundamental way—perhaps even because of that—is a horrific cautionary tale because this noble fellow, this eloquent fellow, this man of moral intentions becomes the most savage of all. And Marlow's fear—that he is like Kurtz, that Kurtz is a kind of version of him—haunts his narrative, and is one of the reasons why he keeps stopping and starting.

But there are other reasons as well, and some of those reasons simply have to do with what we might call the inherent limitations of language. One dimension of *Heart of Darkness* then, and of what I'm calling the drama of the telling in *Heart of Darkness*, is that the drama of the telling takes on what might be called a philosophic or an epistemological dimension. Epistemology is that branch of philosophy that specializes in questions of knowledge and how we know what we know.

And there's a certain sense in which the drama of the telling, this foregrounding of the drama of narrating the story, becomes an epistemological inquiry in which Marlow is testing the limits of language; and not merely the limits of language, but even the limits of our conceptual powers, as if the limits of mind, the limits of the conceptual categories that human beings have to organize the world turn out not to be adequate to the intensity and the terror of the experience of traveling up that Congo River.

So, Marlow's journey up the Congo, a fraught and dangerous journey, is paralleled or recapitulated or re-enacted in the drama of his telling of the story. He lives the drama once 20 years ago when he travels up the Congo, and then he relives it again 20 years later sitting on the deck of that yacht talking to these relatively silent listeners. So when he interrupts himself and attacks his listeners, he's not really angry at them. He's angry at the limits of his own mind, at the limits of his own eloquence, at the limits of his own capacity first of all to understand and then second to articulate what he saw.

All of these kinds of implications are carried, I think, by what I'm calling the drama of the telling. And this drama of the telling also, therefore, reminds us of how textured, how multiformed *Heart of Darkness* itself is, because one can understand Marlow's desperation as a narrator in terms of several dimensions. One dimension we've already talked about—it's a personal quest, right, it's a personal story. It's an autobiographical or private tale. So old Marlow is remembering what happened to him when he was young. And that autobiographical quest, that autobiographical experience, is difficult for him, problematic for him.

But a second dimension of it, related to that, has to do with the way in which Kurtz's example threatens Marlow because he seems to represent a possibility that is resident within Marlow himself.

And a third aspect of it has to do with the way in which Kurtz's example articulates or embodies such a terrifying judgment on the moral and political mission of Europe more generally; because the story's profound indictment of imperialism is much more thoroughgoing, much more powerful than the similar indictment we saw in Kipling's "The Man Who Would Be King," precisely because this philosophic and epistemological dimension is present. It's almost as if Kipling's story stops short of the kinds of insights that Marlow reluctantly, given his strategy of approach and withdrawal as he tells the story, that Marlow finally faces and embraces.

It's as if what Marlow is saying in a certain way is not only that imperialism is a terrible adventure and that the costs, the evils of imperialism involve terrible damage to the native populations, and great, terrible forms of mistreatment and of depredation, really, of foreign peoples and foreign countries. Marlow at one point calls it "a vile scramble for loot." And he understands it in that narrowly political way clearly and fully. But the nature of his narrative also carries further, and this is one of the reasons why Kurtz's association with eloquence, with poetry, with painting, with entrepreneurship is so important, because Kurtz in a certain sense, as I suggested in the last lecture, represents the noblest possibilities of Western culture; and it's the noblest possibilities of Western culture, not it's basest, that lead to the savagery, that end up justifying the savagery, that end up promulgating the worst of the savagery.

So another reason for Marlow's hesitations, for the agony of his drama of the telling, for his reluctance to confront the full implications of Kurtz, which is carried out for us in some respects by his refusal in the beginning to actually get to Kurtz, his postponement in describing the scenes in which he finally meets Kurtz, is a function of the fact that the confrontation with Kurtz, the meeting with Kurtz also nails down these larger philosophic and moral indictments of the culture from which Marlow comes.

So, part of Marlow's near despair at the possibility of communicating what he's seen and known is captured, for me at least, by a great, bitter poem by J.V. Cunningham, a 20th-century poet who used to teach at Brandeis. He's now recently deceased, but he was a very interesting and significant poet. He wrote a terrifyingly powerful short poem called "To the Reader," which captures an aspect of what Marlow is saying or implying in the desperation of his drama of the telling. Here's the poem:

>Time will assuage.
>Time's verses bury

> Margin and page
> In commentary.
>
> For gloss [commentary is gloss] demands
> A gloss annexed [one commentary demands another commentary]
> Till busy hands
> Blot out the text,
>
> And all's coherent.
> Search in this gloss [search in this commentary]
> No text inherent:
> The text was loss.
>
> The gain is gloss.

What a great bitter pun. The gain is gloss, mere glitter, mere surface. So there's a pun on gloss, of course. Gloss means both commentary and a glossy surface.

The implication is that what commentary does is overwhelm experience, cover experience in words so that the original experience is lost, in a sense, in a welter of words, or is uncapturable by those words. So Cunningham's poem seems to me to distill an aspect of Marlow's near despair, an aspect of Marlow's desperation in his telling of the story.

One way we can understand more clearly what is at issue in the psychological, autobiographical, philosophical, and cultural or political implications—all of these dimensions working together in Marlow's drama of the telling and in the adventure itself of *Heart of Darkness*—one way we can understand how these matters are crystallized for us is by talking about the end of the novel.

And the ending of the novel has often puzzled readers, and I'm talking about the scene at the very end where Marlow, having returned to civilization, meets Kurtz's intended—his fiancée. It's a scene that Conrad through Marlow describes with particular clarity and concreteness, and in his typical way the concrete details also carry larger symbolic overtones. He meets the intended in a sepulchral environment, and the implication is that she's living in a kind of tomb, a kind of death in life.

Just before he meets up with the intended, he meets a series of other people who have come to him to try to learn things about Kurtz. A company representative has come to him to try to get Kurtz's papers to see if there's some secret of ivory collection that Kurtz has been hoarding that can help the company. He's visited by a journalist who seems somewhat suspicious of Kurtz and is the man who says that Kurtz should have been a politician. He suggests that Kurtz couldn't write at all, but that he was surely, remarkably, a brilliant speaker and that he could have been a kind of demagogic politician. And then, finally, he meets Kurtz's intended—but before he meets the intended he also encounters someone who represents himself as Kurtz's cousin, and the cousin suggests certain other extraordinary qualities in the dead Kurtz.

And as we know, of course, in the trip back upriver with Kurtz dying on board the steamship, Marlow essentially attends Kurtz alone, and they have a kind of very intense communion in the final days before Kurtz dies. And just before he dies, of course, in the story's most terrible moment in some ways, the most resonant moment, Kurtz says something. He says his final words. He says, "The horror! The horror!"

Well, Marlow goes to Kurtz's intended at the very end of the story still burdened by the meaning of what he has seen, burdened by Kurtz's example, burdened by the horrors that he's witnessed in the heart of darkness, and he finds in Kurtz's intended something terrible and something disturbing. The intended absolutely believes in Kurtz's nobility. The intended is totally devoted to the mythology of Kurtz the great Christian do-gooder, who has gone into the heart of darkness not to make money, not to get wealth, but to transform a savage civilization. And she's utterly committed to the idea that Kurtz's memory must be preserved because of his nobility and remarkableness.

We're told just before Marlow meets with her, for instance, that Kurtz's mother had recently died, and that the intended had been the primary person attending her deathbed. So, she's obviously sort of infiltrated herself into Kurtz's family already. She's made a profession of being Kurtz's fiancée. And it becomes clear as the scene continues that it's going to be her life, that she is sealed in this tomb, this sepulchral environment, forever.

There is something terrible and eerie about it, because we know she's totally deluded about Kurtz. The Kurtz she imagines doesn't exist at all. The actual Kurtz is virtually the opposite, a monster of appetite, a man who took a native mistress, a man who killed at will, a man who put the heads of natives on stakes around his hut as a form of decoration and perhaps to ward off evil spirits; a man who, as Marlow says, had kicked himself loose from the earth, a man who became a god in the land, an embodiment of appetite and savage unrestraint.

So when Marlow encounters the intended and he knows all these things about Kurtz, and is especially, of course, remembering Kurtz's last words, the intended's delusions are horrifying to him, are very troubling to him; and it's a very remarkable scene in which essentially the intended coerces Marlow into reassuring her.

In a letter to his publisher Conrad said that the final scene with Kurtz's intended locked in all the themes of the story, and what happens there has disturbed or disconcerted many, many readers. I've said that the intended coerces Marlow in certain ways, and I have an obligation to mention that this idea of coercion to speak, of conversation that is unmutual, in which one character in some sense bullies or forces or intimidates the other character into saying what he or she wants, is a recurring tactic in Conrad, and it shows Conrad's insight into the idea that conversation itself is a political and coercive, not a mutual, experience.

I'm drawing here on a wonderful book entitled *Coercion to Speak* by the Boston University poet and critic Aaron Fogel, to whom all honor for such a brilliant insight. In Fogel's argument, the archetype of this speech-forcer is Oedipus himself from *Oedipus Tyrannus,* the great Greek tragedy, who forces people in that play to respond to his questions, even though he doesn't realize that the forcing that he's doing is going to lead the reluctant answerers to reveal that he, Oedipus himself, is the very criminal that Oedipus is seeking.

And it is a very powerful sort of archetype for this. In the Conradian versions, the speech-forcer in some of the texts is a torturer. In its most extreme version, in *Nostromo,* for example, there's a man who's threatening torture to someone to try to get him to speak. But in most other versions what happens in Conrad is that the coercion comes from a moral pressure or a psychological pressure that comes from one interlocutor in the conversation.

And the scene of the lie between Kurtz's intended and Marlow is such a scene. And in that scene we see Marlow telling us how he resists giving in to her. But he says, oh, she was so insistent, and finally she says to Marlow, I want to know his last words. I must know his last words. And Marlow says to us, It was too dark. It was too dark altogether. I didn't know what to do. Finally she says again, I must know his last words. And Marlow says, Yes, yes. The last words he spoke were your name. And the intended says, "I knew it!" "I knew it," she says in triumph. And Marlow says, "It [was horribly] dark… [It was] too dark altogether."

And the scene is troubling to us for a number of reasons, but one of the primary ones is that Marlow's whole narration, the whole drama of the telling in the story, has really been about truth and about the difficulty of reaching truth. There's even a moment early in the story where Marlow says explicitly, "I hate … lie[s] … There's a [flavor] of death, a flavor of mortality [about] lies." And yet he lies to Kurtz's intended. He says to her, Kurtz's last words were your name; he spoke your name. And of course that's what she needs to hear; that's what she wants to hear; that's the answer that she has coerced out of him.

And yet there is an ironic and complex way in which Marlow may not have been lying after all, in which the lie he tells to Kurtz's intended may in fact be the truth, because the fact of the matter is, if we reflect on the implications of the story, what we come to realize is that Kurtz's intended believes in an idea of Kurtz as noble and magnificent, as a bringer of light to the places of darkness, as a rationale in a way, as the embodiment of the rationale for the mission of imperialism, for the project of Western imperialism. And from that angle it may turn out, in fact, that Marlow has spoken the truth, because when Kurtz says, "The horror! The horror!" he is describing the intended, isn't he? He is describing the extent to which the intended's beliefs are a kind of horror, and that the moral values and the philosophic implications of the intended's deluded position is a kind of horror.

So there's a wonderfully and disturbingly ironic way in which Marlow's lie speaks a kind of truth, is disturbingly accurate, and isn't after all a lie. No wonder she's confined to a sepulcher, to a chamber of death. No wonder she is, for the reader, a kind of horror.

Lecture Eight
The Shadow-Line—Unheroic Heroes

Scope:

Conrad's *The Shadow-Line* (1917) is one of his last and most compelling works. This lecture examines in detail the plot of this story and its significance. We will look at Conrad's slow, almost frustratingly talkative prologue in which the young seaman is oblivious to the generosity and helpfulness of an older captain. We will then discuss how Conrad both exploits and undermines the conventional myths of heroic adventure on which his story draws. Here, the young hero flounders and makes a series of mistakes resulting from his immaturity and misinterpretation of reality. He eventually succeeds, not by heroic individual effort, but with a good deal of help from unlikely friends. Conrad's heroes discover not their nobility, then, but their human limitations. Lord Jim, like many Conrad protagonists, is "one of us."

Outline

I. Conrad's *The Shadow-Line* (1917), one of his last and most compelling works, was surely inspired in part by his personal life.
  - **A.** Several biographers have suggested that in a period of flagging creativity and deep anxiety, Conrad recovered something of his old powers by returning to the subject matter of his earliest writing.
  - **B.** One source of Conrad's anxiety was that his son Boris was fighting in World War I during the time that Conrad wrote *The Shadow-Line*.
  - **C.** The story has a special interest for college students because it explores the transformation from youth to adulthood.
  - **D.** The book's original title was *First Command*. *The Shadow-Line*, however, is a more ambitious and encompassing title, suggesting that there is no clear demarcation between youth and maturity, perhaps implying that we never complete the process of becoming adult.

II. Let's begin with a general summary.
- **A.** As the story opens, the narrator, a young sailor, quits his job as a third mate on a steam vessel and decides to return to England.
    1. He is baffled himself about this decision, but the novella implies that he is reluctant to choose a profession.
    2. Maturity is about giving up possibilities and narrowing options. Perhaps the young seaman gives up his job because he is not ready to accept the narrowing of selfhood that maturity requires.
- **B.** While he waits in a sailor's home in Singapore for passage to England, the young seaman is sent a message by the Harbor Master, but this message is withheld by the manager of the sailor's home.
    1. The message is an offer of a commission for the protagonist to become captain of a sailing ship.
    2. The young sailor is unaware of the manager's treachery; the latter wants the job to go to another resident whose bill is unpaid.
    3. An older man, Captain Giles, also knows about the message and perseveres against the young hero's rudeness, attempting to alert him to his opportunity.
    4. When Giles finally succeeds, the young seaman accepts his command and realizes that he was destined for a life at sea.
- **C.** The young seaman, now a captain, is impatient about the many mundane tasks required to ready the ship. He believes that once he finally takes to the sea, the problems will be resolved.
- **D.** Once at sea, the young captain experiences even more serious problems: The crew is struck with fever, and the ship is becalmed. The new captain questions his abilities to command.
- **E.** To make things worse, his chief mate, Burns, spreads rumors that the evil spirit of the previous captain is haunting the crew and creating the murderous calm.

- **F.** Eventually, the winds pick up, and with the help of Ransome, his steward, as well as Burns, the young captain steers the ship back to port in Singapore.
- **G.** In the end, everyone survives, and the young captain completes his initiation. In a brief conversation, Captain Giles congratulates the young man for having earned his command.

III. As the above summary implies, *The Shadow-Line* explores one of Conrad's defining themes—that of the double, or *doppelganger*.
- **A.** Another Conrad novella, *The Secret Sharer* (1912), is a parallel or companion to *The Shadow-Line*, and also uses a double-figure to dramatize the maturation of a young captain.
    1. The double of the young captain in *The Shadow-Line* is the weak-hearted cook, Ransome, who quietly offers advice or reassurance whenever the captain doubts himself.
    2. Ransome also prods the captain to do his duty. At a critical moment in the story—the situation on the ship is at its most desperate—the captain is in his cabin, reading his diary, when he looks up and sees Ransome. The captain says, "You think I [should] be on deck?" and Ransome says, "I do, sir."
- **B.** Conrad's doppelgangers can be understood as part of a recurring Conradian drama, a partnership in adventure, in which sailors or other professional comrades band together to survive a catastrophe.
    1. This partnership depicts the human community reduced to its smallest possible unit, endangered but surviving because the characters are able to sustain and support each other.
    2. This test of loyalty, competence, and manhood is at the heart of Conrad's work.

IV. The rich, comic drama in *The Shadow-Line* is evident in the way others try to impart wisdom to the young captain, who continues to misread reality.

A. Giles tries to tell the young captain about the withheld message, but the young captain sees Giles only as a boring old man.

B. The young captain also believes that the home's manager mistreats him because he is young, not because the manager is trying to recoup money owed him.

C. A similar situation occurs with the harbor master and a doctor, further illustrating the young captain's misinterpretation of reality and his obliviousness to the wisdom of his elders.

D. The adventure at sea repeats itself in almost the same way as it had on land; the young captain misreads situations and grows skeptical of Ransome and Burns who are, in actuality, benign characters who end up helping him in essential ways.

E. All of Conrad's characters are orphans, yet one of his great recurring themes is maturity's often useless generosity toward the young.

F. *The Shadow-Line* embodies an anti-heroic vision. The young captain blunders repeatedly and needs help at every stage.

    1. The crew survives and the young captain matures, but he is hardly heroic. He could not have made it on his own. Thus, the myth of heroic individualism is undermined, and Conrad's heroes discover the limitations of being human.

G. D. H. Lawrence had contempt for Conrad's quavering heroes, but the Conrad protagonist, in his weakness and imperfection, as we're told in *Lord Jim*, is always one of us. When Ransome leaves the ship at the end, the young captain listens to him walking up the stairs, fearing for his weak heart, "our common enemy it was his hard fate to carry consciously within his faithful breast." We are all human and, therefore, faint-hearted.

Essential Reading:
Conrad, *The Shadow-Line: A Confession*.

Supplementary Reading:
Conrad, *The Secret Sharer*.

Thorburn, *Conrad's Romanticism*, pp. 128–152.

Questions to Consider:
1. What is the relation between the extended "prologue" to the story that takes place in the sailors' home and the actual adventure at sea in the second part of the text?
2. What does the story seem to say about preconceived ideas concerning heroic individualism and actual experience?
3. What is the significance of the story's title?

Lecture Eight—Transcript
The Shadow-Line—Unheroic Heroes

Heart of Darkness is an early masterpiece. *The Shadow Line*, the subject of today's lecture, was published in 1917, and it's Conrad's last compelling work of art. Several biographers have suggested, in fact, that in a period of flagging creativity and deep anxiety, Conrad recovered something of his old powers by returning to the subject matter of his earliest work. One source of his anxiety was that his son Boris was fighting in the First World War during the time of the story's composition.

The story has a special value, a special interest, a special resonance for undergraduate students, I have found, and I think that one reason for this is that even though very few of the students I've taught in my career have an ambition to become sailors, they respond with particular intensity and sympathy to the story's recognition that the transition from youth to adulthood involves, among other things, an anxiety about whether a young person is capable of handling responsibilities.

Conrad's novella is deeply about this in a series of ways, and there are a series of encounters between the young narrator, the young protagonist of the story, and older figures, as I will describe in some detail later in this lecture. And it is also partly a dramatization of a young man's experience in taking command of a sailing vessel in which most of the people he's commanding are older than he is. And I think this in particular has appealed to my MIT students, who both look forward to but also fear the likelihood that when they graduate from MIT they'll be put in positions of authority that give them responsibility over men and women older than they. And this is an issue that Conrad's novella deals with with some power and some directness.

Let's begin with a plot summary. "Only the young have such moments," Conrad says in the very beginning, and it is a deeply autobiographical story. Its original title was "First Command," and this title is a good title, I think. It would certainly work for this novella. However, I think we come to realize that "The Shadow Line" is an even more resonant and powerful title. I think one reason for that is that the story dramatizes something that's at the heart of Conrad's anti-heroic vision of the world.

One aspect of that vision is that there is no single decisive moment in which one goes from being a child to being a man, no one clear line of demarcation in which one stops being immature and suddenly becomes mature. Rather, one grows into maturity slowly over a period of time. The line is a shadow line; it's a blurred one, and indeed, not only is there no exact moment of transition, it may be that the final endpoint is never fully reached. At the very end of this story there is even a hint of a suggestion, in a kind of irritation that the young narrator shows toward his older mentor, Captain Giles, that he himself may not have fully become totally mature yet, even though he has gained a tremendous amount of understanding and wisdom in the course of the ordeal he faces.

The plot itself is a relatively straightforward and simple one. The young protagonist of the novella, for no apparent reason, gives up his job as a third mate on a steam vessel, and decides to give up his career of the sea and to go home to England. He takes lodgings in the sailor's home in Singapore. He's mystified himself, baffled himself, about why he has done this. And it isn't until later in the story that we come to understand that one of the reasons may well have been that he was approaching the point in which one has to make final decisions about one's vocation; and once one chooses to be a captain as against being a mathematician or an astronomer or a fisherman, one is closing off options. And the implication of the story is in part that one cost of maturity is the giving up of possibilities.

So, one suggestion that the story makes at the very beginning is that the young man's bafflement over why he is so impatient, and throws over this job, is that he is reluctant to accept the narrowing, the definition of selfhood that comes from growing up, that comes from maturity.

While he's in the sailor's home a message comes for him, but the manager of the sailor's home delays giving it to him because he would prefer that the message be given to someone else who owes him money who is also renting a room in the sailor's home. The message is from the harbormaster's office, and it contains a summons to the harbormaster's office, with the implication that the person who responds to this message will be given a command of a vessel.

The treachery of the manager of the sailor's home is something to which the young captain is oblivious, and a good part of the first section of the novella is given over to a comic drama in which an old captain, Captain Giles, tries at first very unsuccessfully to get the young man, the young narrator, to see that the message is for him. Finally he does succeed and the young man goes to this harbormaster's office, is offered his first command, and proceeds to his ship to take over that command.

When he's on some sort of a sailing vessel that's taking him to the outer part of the harbor where the ship is moored, he begins to realize that he was born for the sea, and he begins to realize that had he lost his opportunity to become a captain it would have been a catastrophe for him in a personal and in a professional sense.

After he assumes his command he has to go through a series of very elaborate and in many ways tedious preparations, with which he's incredibly impatient. Some of the crew has been ill and he has to take care of their illness. He has to check the ship's supplies. And he keeps saying to himself, Going off to sea would be "the only remedy [to all our difficulties]." He's very impatient to get to sea.

Finally, he gets to sea, and as you might imagine, given the fact that this is a Conradian initiation story, once he gets to sea instead of having all his difficulties solved he is only now confronting serious difficulties. After a short time the ship is becalmed, the entire crew is struck down by fever, the young captain begins to discover a sense of unreadiness and even unworthiness in himself and isn't sure whether he has the gumption, the intelligence, the capacity to run the crew.

The ship is stuck in the middle of the ocean. It's a sailing vessel, so it can't get back to port without wind, and it looks as if the entire crew will perish. Winds do in the end come up, and the young captain, with the help of his steward (a man who had signed on not as an able-bodied seaman, but as a man who would not be required to do physical labor because he has a bad heart), a man named Ransome, helps him to steer the ship back to port in a very heroic and remarkable way.

Let's listen to the very first sentence of the novella:

> Only the young have such moments. I don't mean the very young. No. The very young have, properly speaking, no moments. It is the privilege of early youth to live in advance

> of its days in all the beautiful continuity of hope which knows no pauses and no introspection.

One closes behind one the little gate of mere boyishness—and enters on an enchanted garden …

> … And the time, too, goes on—till one perceives ahead a shadow-line warning one that the region of early youth, too, must be left behind.

Finally in the story, a great storm comes up, and although almost the entire crew is struck down by fever, the young captain himself was himself healthy, and his steward, a man who had signed on the ship not as an able-bodied seaman but essentially as a valet and as a cook because he has a bad heart, and is unable to do strenuous exercise. This man's name is Ransome. And Ransome helps the young captain sail the vessel back into port in a final desperate heroic act. Ransome acts like the strongest of able-bodied seamen, climbing the rigging and doing all kinds of astonishing physical feats, risking his life with every gesture because he has a weak heart; and the young captain and Ransome manage to get the vessel back to port. They never get where they're going. They turn the vessel around and they go back to Singapore from whence they'd left.

But it is a kind of heroic survival tale, and they arrive back in Singapore safely. The crew is taken off the ship. Everyone survives well, and the young captain has had his difficult initiation.

There's a final scene at the end of the story in which old Captain Giles, who had helped the young man gain his first command, has a brief conversation with him. It's a kind of coda in which in a certain sense the old captain is blessing the younger one and saying, you've now made it. You're now a part of the fraternity of captains. You've now earned your right to call yourself a seaman, a captain, a commander.

Well, here's how the story begins. Listen to the very opening paragraphs:

> Only the young have such moments. I don't mean the very young. No. The very young have, properly speaking, no moments. It is the privilege of early youth to live in advance of its days in all the beautiful continuity of hope which knows no pauses and no introspection.

> One closes behind one the little gate of mere boyishness—and enters on an enchanted garden. ...
>
> ... And the time, too, goes on—till one perceives ahead a shadow-line warning one that the region of early youth, too, must be left behind.

And I think by the time we finish the story, we realize that the whole voyage has been his shadow line. And it's a shadow line, I think we realize, because there's not one single moment which demarcates the second in which one stops being a young man and becomes an adult, or in which one stops being immature and becomes mature, but rather that it's a very gradual process that may not in fact complete itself even by the end of this tale.

Well, contrast that quiet narrative voice that I've just read to you with the desperate, driven tone of Marlow in *Heart of Darkness* or the Marlow we hear in Conrad's great novel written around the same time, *Lord Jim*. The drama of the telling—what I called the "drama of the telling" in the last lecture: that story which displaces or competes with the traditional narrative matter of the story, the drama of the narration, the drama that tells the story of the difficulty of telling the story—is in *The Shadow Line* highly muted. It's there; it's present; we can hear it, but it's much less powerful, much less central to our experience of the story.

This narrator is an older narrator, and we can hear the drama of the telling in the amused irony and distance of the older narrator's voice. He's older than the young man he describes. Maybe he's retired. He's looking back (like the narrator in the first three stories of Joyce's *Dubliners*) at his youthful adventures with a mixture of embarrassment and maybe with some surprise.

Now, there's another, even more powerful connection between this late novella and earlier Conrad, and that has to do with Conrad's preoccupation with what has been called the theme of the double, or the doppelganger. Like certain other writers—like Robert Louis Stevenson, who wrote a very famous double novel, *Dr. Jekyll and Mr. Hyde*; like Dostoyevsky, who wrote a magnificently complex novella called *The Double*—Conrad is very interested in what we might almost think of as split personalities; in the way in which the personality or the self itself may be understood to have a fundamentally divided nature.

And in some of Conrad's richest stories and novels, this doubling takes on a profound psychological and moral resonance. One powerful instance of it is something we've already talked about. It has to do with the way in which, in *Heart of Darkness*, Kurtz is a kind of demonic double, a kind of murderous counterpart to Marlow himself; a murderous counterpart to Western civilization itself, in some respect.

And there's another story of Conrad's, called *The Secret Sharer*, which is in one sense a kind of counterpart tale to *The Shadow Line*, as if they are twin stories. *The Shadow Line* might be said to be the double of *The Secret Sharer*. In *The Secret Sharer*, written much earlier than *The Shadow Line*, Conrad also tells the story of his first command, and dramatizes his own sense of nervousness and unreadiness at the responsibilities of adulthood.

In *The Secret Sharer* Conrad's psychological or psychiatric interests are more explicit, and in that story the double of the young captain is also a captain newly in command in *The Secret Sharer*. And at a certain point a strange man, swimming alone in the sea, comes up out of the sea, climbs up the Jacob's ladder on the side of the ship in the dead of the night; and it just so happens that the young captain is himself on watch, and brings the young man in, and he's an escapee from another ship. He's actually wanted for murder. And the young captain protects him. He's actually the same size as the young captain, and he can wear the same clothes as the young captain. They have a similar background.

And Conrad's very explicit about the extent to which in *The Secret Sharer* this double is a kind of counterpart, a kind of outlaw version of the young narrator. And *The Secret Sharer* is in some sense, I suppose, about the need to purge oneself or free oneself not only from one's sense of youthful unreadiness for the responsibilities of adulthood, but also of purging oneself of impulses toward lawlessness and outlawry that perhaps are also part of early youth.

In *The Shadow Line*, Conrad takes a more, we might say conservative, but in any case a much more mature and I think generous view of these tendencies, sees them in less psychiatric terms; but in *The Shadow Line*, too, there is a profound and interesting treatment of the double theme, and in *The Shadow Line* the duplicate or the double of the young narrator is the weak-hearted steward, Ransome. Ransome is in fear of dying every moment that he is on board the ship because of his weak heart, and the young captain doesn't fully understand the degree to which Ransome is useful or helpful to him until the story plays itself out.

But in the wonderful complex drama that's acted out in the ship's aborted and balked voyage out into the Gulf of Siam, and then back to Singapore, what we see again and again is the young narrator doubting himself, hiding in his cabin, writing in his diary: I think I might not be ready for these problems. I always wondered. "I always [thought] I might be no good," he says at one point. And now I'm failing it. Now I'm failing. And in almost every moment in which the young narrator, the young captain, feels these doubts, Ransome appears beside him, or appears nearby, and very quietly, without nagging or without seeming to give him orders, seems to sort of provide a kind of reassuring presence.

And at one point the young narrator, when the ship is in its most desperate moments—they'd been becalmed for a long time; all the crew is ill—Ransome appears at his side and the young narrator is in his cabin looking down at his diary in which he's confessed his failings, and he looks up and he sees Ransome, and he says, "You think I [should] be on deck?" And Ransome says, "I do, sir." And it's a critical moment in the story. And then he goes back up on deck. And of course Ransome does perform heroically on the voyage back to Singapore.

So a defining element in Conrad's work, a fundamental aspect of his interest in these stories of initiation, of growth, of maturation, involves this treatment of the double. And in *The Shadow Line* we might say he gives his most mature and least troubled version of this tale.

One way to describe what I'm talking about is to say that Conrad again and again in his novels, and especially powerfully in *The Shadow Line*, creates what could be called "a partnership in adventure" in which two unequal characters—they're either unequal because one is older and much more experienced than the younger one, or because one is damaged or hurt in some way, as Ransome is in *The Shadow Line*—an adventure partnership in which two characters of unequal qualities band together in order to survive some catastrophe.

And it's one of Conrad's most radical visions of human community; human community reduced to its smallest possible unit, endangered but not quite destroyed, surviving because the characters are able to sustain one another and to support one another in their ordeal. Well this adventure partnership, this test of loyalty, competence, and manhood is at the heart of Conrad's work, and is at the very heart of *The Shadow Line*.

One of the most interesting aspects of *The Shadow Line*, which has puzzled and troubled some readers, is its apparently strange proportions; and it's worth emphasizing these strange proportions because they throw a special light on what I've been calling "the adventure partnership." The most interesting aspect of the difficult structure of the story is that there is a prologue in the sailor's home, that episode I've already described to you, that lasts almost as long as the adventure at sea itself. And it's mostly dialogue. And it takes place only over a very short time—a day or a few days. And much of it has to do with dialogue in which the young captain describes his impatience and annoyance with boring, old, garrulous Captain Giles, who doesn't seem to want to shut up.

And of course what's going on there, although the narrator doesn't at first realize it, is that Captain Giles, despite the fact that the young narrator is being disrespectful to him, is being nasty in some ways to him, is certainly not showing him an adequate regard, that nonetheless Captain Giles persists and persists in trying to make the young narrator understand that a very important message has arrived for him. And the young narrator is such an arrogant and such a frightened young man that he doesn't allow himself to see it at first.

There's an earlier prefiguration of this behavior when the narrator, the young man, first arrives at the sailor's home, the steward who runs the sailor's home asks him for payment up front, and the young narrator gets outraged by this. He says, How dare you ask me that, and he refuses to pay. And of course he's angry because he believes that he's being mistreated because he's so young. You would never ask an older man an established person this; how dare you ask me.

Of course, as with many things that happen in this story, the young man is misreading reality out of his youthful inexperience; because as we realize later, and as he realizes later, the steward is asking him for the money because he has someone else in the sailor's home who hasn't been paying the rent. He has no intention of insulting the young narrator, and that wasn't even in his mind, but the young narrator misreads appearances out of his youthfulness and out of his inexperience.

And he does the same thing again with Captain Giles. He misreads Captain Giles's benign intentions for him by being oblivious to the hints that Captain Giles keeps trying to give him. Finally, Captain Giles gets through to him, and the young man rushes to the harbormaster's office, receives his first command. The scene in the harbormaster's office is in some ways also another enactment of this recurring Conradian drama in which an older man tries to pass on his wisdom to a younger and often reluctant younger man, because the harbormaster is another kind of senior figure, another kind of surrogate father figure who interrogates the young man to make sure that he's competent and capable of taking on the command. And the primary reason that the young captain is given the job is that the previous captain he had worked for on that steam vessel, whose job he had thrown over for no good reason, had given him a very good reference, and had said, "This is a competent seaman."

So again and again we see enacted in this story of Conrad's the young man being helped and profoundly aided by sometimes older people, sometimes younger men like Ransome, who offer their aid almost despite the young man, certainly in the face of indifference and sometimes even disrespect on the part of the young man.

Though all his people are orphans—Conrad never wrote a story in which there were fathers and sons—Conrad remains one of the great portrayers of the anguished impotence of fatherhood. One of his great recurring subjects is maturity's often useless generosity toward the young.

Well, the adventure that the young man engages in when he sets out to sea repeats in almost exact proportions the trivial or comic adventure that he had enacted in the sailor's home, an adventure in which he misreads reality, makes all kinds of mistakes about what he sees before him, and is helped along his way by benign and helpful others. Some of the mistakes he makes involve, for example, his idea that the captain who preceded him must be an admirable character; and because it never occurs to him that the captain he has replaced might have been a treacherous or an evil person, he does not check the quinine himself to make sure that there really is quinine in those bottles. Had he done so before he left port. And there is a scene in the novella in which we see him conferring with a port doctor, and the doctor and the young captain talk a good deal about the stores on the ship and so forth, and the doctor warns him that although his men have been cured of their worst fever, if he ships out too quickly he may find that the fever will return.

But the young captain keeps thinking, Well, if only I could get out to sea everything would be fine. But once he's given the command, he has all kinds of minor difficulties in outfitting the ship and getting ready to leave, and he's incredibly impatient to get out to sea. His impatience is part of the reason that, when he does get out to sea, he's unready for the difficulties that he faces. And once he's out to sea he opens the drawer in his cabin and he finds a letter there from this doctor, another wise elder who had wanted to help him, and this letter from the doctor says, I didn't want to discourage you because you were so excited when you were leaving port, but let me remind you that your men may come down with fever. Don't worry too much about it. The quinine will see you through. And the implication again is if he had checked the quinine, a lot of the difficulties that he faces when he's out there on the ocean would have been avoided.

So he misjudges the prior captain, thinks that he's a reliable man when he's not a reliable man. He misjudges Captain Giles, underestimates him. He misjudges Ransome, whom he takes for granted for most of the voyage until he comes to realize that Ransome has been utterly essential to the modest success of the voyage.

And there's another man on board the ship with whom he has some difficulties, and about whom he's made a terrible mistake, and this is a character named Burns. Burns is older than the young narrator, and he was the first mate on the ship, and there's some suggestion that Burns is resentful that the young captain has been given the command instead of him. He's also very feverish and sick before the voyage even begins, and he begs the young captain to take him with him on the voyage because he needs the money, and the young captain is advised by the doctor not to take Burns with him; but he is a softie, and he allows himself to accept the appeal that Burns makes.

And once they're out to sea Burns doesn't become a horrific problem, and in fact at the very end of the story when the ship makes its wild desperate run into port, Burns is actually helpful, because although he's struck down by fever—he's incredibly sick—he straps himself to the wheel of the ship and steers the vessel while the captain and Ransome climb the rigging and keep the ship afloat in the great hurricane that runs the ship back into Singapore. So Burns's example is an ambiguous one.

But Burns also spreads rumors about the old captain being an evil spirit who's haunting the ship and who is responsible for the ship being becalmed, and he rattles the crew; and he certainly disturbs the young captain, who is very upset by Burns's constant suggestions that his superstition is the explanation for why the ship has been becalmed.

So again, he misjudges Burns; he misjudges Ransome; he misjudges the events around him in ways that suggest youthfulness, inexperience, something of the arrogance of youth. Throughout the tale, then, the young man misjudges men, misjudges events in ways that had he been left on his own would have brought him low. But he is not brought low, because he gets a lot of help from his friends, and from friends he doesn't even recognize as friends. What this reminds us of, and why this tale is such a powerful and important one I think, is that it embodies perhaps more fully and more austerely than any other Conrad story or novel what we might call his deeply anti-heroic vision.

The crew survives in this story. No one dies. The young hero does mature and becomes something close to an adult before the final words of the story. But this is hardly a heroic beginning to the young captain's career. He has blundered and nearly undone himself repeatedly; first with Giles, then once aboard ship, in a whole series of ways. At every stage the hero needs help, couldn't have made it on his own. So the myth of heroic individualism, which underlies the traditional adventure story, underlies the kind of story that Robert Louis Stevenson, for example, was publishing very widely in the years when Conrad was first thinking about becoming a writer, and were stories that were surely known to Conrad and influenced Conrad, that myth of heroic individualism is undermined in Conrad's story. In the Stevensonian adventure story the young hero was released into an exotic space, and what happens there is that he discovers how powerful, how heroic, how remarkable he is. What Conrad's heroes discover when they go into those exotic spaces, when they take their adventures on the high seas, are the sure and terrible limitations of their humanity.

So it's a failed voyage; a balked and stumbling maturation. Conrad is suggesting that we rarely choose our solidarities with the past and the present and often are dragged into them screaming. It's worth recalling D.H. Lawrence's angry contempt for Conrad here as we close, because he disliked Conrad's quavering heroes, and in a wonderful line he said: "Snivel in a wet hanky like Lord Jim," said Lawrence.

But the Conrad protagonist, in his weakness and imperfection, as we're told in *Lord Jim*, is "always one of us." When Ransome leaves the ship at the end, the young captain listens to him walking up the companion stairs, fearful of his weak heart, described as "our common enemy it was his hard fate to carry consciously within his … breast." In Conrad's novels and stories, we are all human, and therefore faint-hearted.

Lecture Nine
The Good Soldier—The Limits of Irony

Scope:

Ford Madox Ford's life was complicated with personal scandal and damaged to some extent by his own boastfulness, but Ford would be owed more than a footnote in literary history even if he had never written a word himself. He was a genius and had a talent for recognizing genius in other writers. His collaborations with Conrad had critical importance for both writers. The men devised a theory of "impressionism" that clarified and justified the wayward, non-chronological structure and unreliable narrators of their works. *The Good Soldier* is Ford's masterpiece—a book so drenched in irony and indirection it may be said to carry one principle of Modernism (that of the unreliable narrator) almost as far as it can go. How Ford incites the reader to recognize, understand, and evaluate the narrator's evasions and self-deceptions is a central question, whose full answer leads us to see more deeply into the book's terrible drama of sexual shame and the coercions of religion and social propriety.

Outline

I. Ford Madox Ford (1873–1924) was a prolific and influential writer and editor.

   A. He wrote more than 80 books [sources vary on the number], including 32 novels, some of which were collaborations with Joseph Conrad. His most compelling fiction, in addition to *The Good Soldier* (1915), is a tetralogy of novels collectively called *Parade's End* (1924–1928).

   B. Other works include *Joseph Conrad: A Personal Remembrance* (1924), published in the year of Conrad's death.

      1. Ford claimed an intimacy with Conrad that Conrad's friends and family resented, believing the claims unjustified.

      2. They were mistaken, however, as the two men's letters and collaborative works prove.

- C. Ford's *Portraits from Life* (1937) contained revealing tales about such writers as Henry James, Stephen Crane, D. H. Lawrence, Thomas Hardy, H. G. Wells, and Theodore Dreiser.
  1. Ford claimed to be partly responsible for the success of these writers.
  2. There is some truth to his claims; Ford was a great editor and discoverer of genius.
  3. He was founder and editor of *The English Review* in 1908–1909, which discovered D. H. Lawrence and also published work by Hardy, James, Galsworthy, Bennett, Conrad, and Wells.
  4. In 1924, Ford became editor of *The Transatlantic Review*, which published Joyce, Pound, Stein, Cummings, and Hemingway.
  5. Some of the writers Ford helped—notably Hemingway—wrote condescendingly about him.
- D. Ford's scandalous divorce from his first wife led to a series of affairs with women who wrote about him, including Jean Rhys (*Postures*, 1928; published in America under the title *Quartet* in 1929).

II. Conrad and Ford collaborated on three works and developed a theory about how Modernist fiction should be written.
  - A. Although a generation younger than Conrad, Ford collaborated with him in the formative first phase of Conrad's career, a collaboration that had critical importance for both.
  - B. In Ford's memoir of Conrad, he describes their "impressionism," as he called it.
    1. "We agreed that the general effect of a novel must be the general effect that life makes on mankind. A novel must therefore not be a narration, a report," Ford wrote.
    2. In other words, life is learned accretively, in fragments; it is not laid out for us in expository fashion. Conrad believed that Modernist fiction should be written to reflect this. It could not speak to the reader in the way that traditional 19th-century novels did.
  - C. Their collaborations include *The Inheritors* (1901), *The Nature of a Crime* (in *The English Review* in 1906; in book

form 1924), and *Romance* (1903), which was a full-fledged, first-person adventure tale developed from Ford's own "Seraphina," a precursor to Conrad's *Nostromo* (1904).

1. He was founder and editor of *The English Review* in 1908–1909, which discovered D. H. Lawrence and also published work by Hardy, James, Galsworthy, Bennett, Conrad, and Wells.
2. In 1924, Ford became editor of *The Transatlantic Review*, which published Joyce, Pound, Stein, Cummings, and Hemingway.

D. Conrad devoured Ford's work but also took advantage of him. He then distanced himself from Ford following the latter's divorce and extra-marital exploits.

III. Ford published *The Good Soldier* in 1915 when he was 42 years old.

A. The story centers on two couples, John and Florence Dowell and Edward and Leonora Ashburnham. It is told by John Dowell, who opens with, "This is the saddest story I have ever heard."

B. The Dowells move to Europe because Florence has always dreamed of traveling there. In Europe, they meet the Ashburnhams, who appear to be a happy, normal couple.

C. After nine years of friendship, however, John Dowell discovers that Edward and Florence have been having an affair, and that each had been unfaithful to their spouses earlier.

IV. *The Good Soldier* is so drenched in irony and indirection that it may be said to carry the Modernist principle of the unreliable narrator as far as it can go.

A. Recognizing the irony is a central and problematic aspect of Modernism itself. Even 90 years after the book's publication, critics and readers still argue about how to understand Ford's narrator, John Dowell.

B. Dowell's voice has no omniscience or authority; he is seductive yet apparently guileless, candid, and direct.

- **C.** Dowell's discourse is also full of contradictions, false dichotomies, confessions of ignorance and helplessness, and incessant rhetorical questions.
- **D.** The opening pages show attentive readers that the narrator cannot be trusted, as evidenced in his summation of his marriage to Florence as a "beautiful minuet" and immediately after that as "a prison full of screaming hysterics."
    1. One moment, Dowell says that Edward Ashburnham is noble and a good soldier; the next, he describes Edward as a sentimental fool ashamed of his carnal desires.
    2. One moment, he calls his wife "poor, dear Florence"; the next, he calls her a manipulating creature he resents.
- **E.** The story's sordid details seem to emerge without Dowell's full understanding, almost as if he lets them drop by accident. The interrogative is Dowell's favorite mood, and his narration is full of questions, tentative and passive.
- **F.** Later in the story, Dowell says, "I have, I am aware, told this story in a very rambling way so that it may be difficult for anyone to find [a] path through what may be a sort of maze. I cannot help it."
    1. The narrator's constant confessions of blindness, inadequacy, and ignorance are the key to the novel and to the particular nature of Ford's drama of the telling.
    2. A separate narrative parallels and even dominates the traditional story, just as in *Heart of Darkness*. *The Good Soldier* is told in a digressive, non-chronological fashion in which major events are disclosed reluctantly, often obliquely.

V. Modernist texts demand collaboration from the reader; this interactivity is complex and morally challenging.
- **A.** Readers of *The Good Soldier* have the difficult task of reconfiguring the story's chronology and comparing Dowell's own reconstruction of the events to his judgments and opinions.
- **B.** This is a more strenuous aesthetic and moral experience, in some respects, than that demanded by more traditional novels.

- **C.** Readers are expected to collaborate in the interpretation of the events and, to some degree, their outcome.
- **D.** Henry James believed that the omniscient narrator of the 19th century betrayed the extent to which the novel was a contrived artifact.
  1. James used a "center of consciousness," through which the story was narrated.
  2. For James, this technique was critical for a sense of believability.
  3. Conrad and Ford adapted this concept in their theory of impressionism.
- **E.** In *The Good Soldier*, the narrator's puzzled, contradictory voice and the circling shape of his story express not a desire to understand but to evade understanding.
  1. The drama of the telling is a psychological stratagem in a desperate effort at self-justification by a damaged, morally obtuse man whose wife and friends have betrayed him to his face for nine years.
  2. Dowell tells his story to justify his life and to pretend that he could not have known about the philandering. Eventually, the reader realizes that the recounting of the story is an act of pathological self-deception.
- **F.** Comparing Dowell to Conrad's Marlow, we see that Marlow is not disturbed in the sense that Dowell is, and that he is aware of the limitations of language and memory. Conrad is committed to a rhetoric of uncertainty. In contrast, Dowell's is a rhetoric of impotence, of helplessness.
- **G.** If we want to distill a stylistic and moral essence on a similar axis for other Modernist writers, we might describe Henry James's characteristic tone as a rhetoric of qualification; D.H. Lawrence, the rhetoric of incantation; Joyce, inclusion; Faulkner, conjecture and speculation.

Essential Reading:

Ford, *The Good Soldier*.

Schorer, "An Interpretation" (introduction to the Vintage edition of *The Good Soldier*).

Supplementary Reading:

Ford, *Some Do Not* and *No More Parades* (vols. 1 and 2 of *Parade's End*).

Questions to Consider:
1. What features of the narrator's vocabulary and tone of voice are decisive in suggesting that his view of events is not trustworthy?
2. What is the consequence for the reader in recognizing that the narrator is unreliable?

Lecture Nine—Transcript
The Good Soldier—The Limits of Irony

Ford Madox Ford is the least well known of the writers in this course. He has perhaps brought that fate on himself, as I'll explain in a moment. But he deserves a wider fame.

We might begin by saying a few words about his prolific and, in some ways, scandalous life. He was the author of more than 80 books, 32 novels including collaborations with Conrad. His major fiction includes *The Good Soldier*, the text we're going to be talking about, and a wonderful tetralogy of novels, four linked novels written in the period 1924–1928 under the broad title *Parade's End*. This tetralogy, I think, in its own way rivals the other major English fiction of the period, including Virginia Woolf's major novels and Joyce's *Ulysses*; and like those novels it is preoccupied with what we might call the transition from the 19th to the 20th century, the transition to modernity.

Parade's End is a remarkable series of books, and I urge those of you who find *The Good Soldier* interesting to take up those four novels as well. They're set during the First World War and they tell a story in a certain sense that's similar to the more oblique version of this tale that's told in *The Good Soldier*, of the breakdown of values, the breakdown of belief systems that might be said to define the before and after story of the First World War.

He's the author of many other interesting works as well, among which I would single out especially a wonderful personal book called *Joseph Conrad: A Personal Remembrance*, published in 1924, the year of Conrad's death. The book came under some cloud because, in the book, Ford rightly claimed a great intimacy with Conrad. He had collaborated with Conrad on several books. But there was already by 1924 a collection of Conrad idolaters, led, in fact, by Conrad's widow, Jessie Conrad; and they were very resentful of the fact that Ford was claiming an intimacy they felt was unjustified. They were mistaken about this, as the collaborations between Ford and Conrad themselves show, as well as certain letters that were exchanged between Conrad and Ford in the period of their collaboration which have now been made available to scholars.

Ford is also the author of other interesting work. One of these books is a book called *Portraits from Life*, published in 1937, and it includes sketches of almost every famous writer in the English language from the period in which Ford lived: James, Stephen Crane, Conrad, D. H. Lawrence, Thomas Hardy, H. G. Wells, Dreiser, other writers, all friends or acquaintances of Ford, and in *Portraits from Life* Ford paints portraits of these writers.

The book is full of interesting anecdotes and revealing tales about his figures, but it also suffers from a weakness that helps to explain why Ford's reputation remained under a cloud for such a long time. Because although his book is very revealing, he also harmed himself terribly by writing so much and so extravagantly about his acquaintances with these writers. In every one of these accounts Ford comes across as the man who saved a particular writer from obscurity, or discovered a particular writer, or was responsible for improving a particular writer's work; and although there's a great deal of truth in the anecdotes that Ford tells, it's also impossible not to feel an element of exaggeration and boastfulness in them. It's a sad thing, in a way, because Ford certainly didn't need to boast; but by 1937 he was not a well-known writer any longer and he felt, I think rightly, that he had produced major work that was now being ignored.

One of my favorite titles of Ford's tells something about the kind of person he was, his ambition and his sense of ownership of literature. The title, published in 1938, is *The March of Literature from Confucius to the Present Day*. And Ford is one of the very few people who would claim expertise across such a range of literature. He described himself in his last years when he was teaching in the United States in the Midwest at Olivet College as an old man mad about writing.

Another factor in the scandal of Ford's life had to do literally with the fact that he was the source of scandal. He received a tabloid divorce from his first wife, and it was covered in all the London newspapers, to the great embarrassment of Ford's friends, including Joseph Conrad, who distanced himself from Ford when the unsavory publications began to appear.

And he then conducted a series of affairs with other women, often very gifted women, who wrote about him obsessively after he left them. One of the most famous of these is Jean Rhys, the gifted novelist best known for the novel *Wide Sargasso Sea*; but her gimlet-eyed story of her affair with Ford Madox Ford, as always with Rhys a mistress's fable, is titled *Postures*. It was published in 1928, and in the following year, 1929, in America, under the title *Quartet.*

The American poet Robert Lowell, who worked as Ford's personal secretary for a few months, caught the conflicting qualities in Ford—his boastfulness, but also his touch of genius—in a memorable poem that was published in Lowell's book, *Life Studies*. Here are some lines from it:

> But master, mammoth mumbler, tell me why
> the bales of your leftover novels buy
> less than a bandage for your gouty foot.
> Wheel-horse, O unforgetting elephant,
> …. I'm selling short
> your lies that made the great your equals. Ford,
> you were a kind man and you died in want.

Ford's friendship with Conrad is worth a further note. Although he was a generation younger than Conrad, he collaborated with the older writer in the formative phase of Conrad's career. Their collaboration is still rarely studied, although it was a very important experience for Conrad whose mastery of English at this stage in his life was shaky.

And in the memoir I mentioned called *Joseph Conrad: A Personal Remembrance*, Ford describes their impressionism, what he calls the "impressionism" (using that term) that he and Conrad worked out as their theory of how fiction should be written. He described that quality, those principles, in ways that clarify and justify the wayward, non-chronological structure and the unreliable narrators of both Ford's and Conrad's great novels. The very essence of Ford's idea in *Joseph Conrad: A Personal Remembrance* of this impressionism is essentially that a novel cannot speak to you in the way that the traditional 19th century novels do.

In a famous sentence, Ford wrote: "We agreed that the general effect of [the] novel must be the general effect that life makes on mankind. A novel must therefore not be a narration [or] a report." And he went on to explain how when you learn anything in life, you learn it accretively, you learn it abortively. You learn it in fragments. It isn't as if life presents you with expository paragraphs setting the stage and explaining everything for you, and Ford goes on in this remarkable passage to describe essentially the structure and the form of both Conrad and Ford's greatest novels.

Their collaboration covered three books: a book called *The Inheritors,* published in 1901, which was a kind of science fiction novel; a very short novella or long story called *The Nature of a Crime,* written in 1906 but published in the *English Review* in 1909, and then published in book form in the year of Conrad's death, probably to capitalize on Conrad's name; and then their most important book, a book published in 1903 called *Romance.* This last, *Romance,* is a full-fledged first-person adventure tale with a Stevensonian narrator-hero, developed from a manuscript by Ford, originally titled "Seraphina." It's a precursor to Conrad's great novel published in 1904 called *Nostromo,* and bears all kinds of resemblances to *Nostromo.* It's my belief that, in a certain sense, Conrad went to school on "Seraphina," later called *Romance,* and developed many of the strategies and tactics that he used to such great effect in *Nostromo* itself.

In one sense one might say, and I believe this is true, that Conrad devoured Ford, used Ford, took advantage of Ford, and then when Ford's reputation declined and his divorce got in the newspapers, Conrad backed away from him. It's an unsavory aspect, I think, of Conrad's life, that he never showed himself more grateful to the young writer who shared so much of his life and so much of his work with him in the formative stages of Conrad's career.

Ford was also a great editor and a discoverer of genius, and he would be owed more than a footnote in literary history even if he'd never written a word himself. In the period 1908–1909 he was the founder and editor of one of the most distinguished literary magazines ever created in the English language. It was called *The English Review,* and it was the magazine that first published D. H. Lawrence. It published Hardy, Henry James, Galsworthy, Arnold Bennett, Conrad, H. G. Wells. It was an immensely distinguished and remarkable literary journal.

And then, a generation later, Ford became the editor of *another* magazine that introduced *another* generation, the Modernist generation, especially certain American Modernists, to the world. This magazine was called the *Transatlantic Review,* and it published work by Joyce, by Ezra Pound, by Gertrude Stein, by E. E. Cummings, by Hemingway.

One anecdote that I find irresistible and want to share with you has to do with Hemingway's relation to Ford, and I guess I think it's important because it crystallizes something of the way in which Ford has suffered mistreatment at the hands of people he helped. Hemingway was in many ways very ungrateful to Ford. Ford had been very generous to Hemingway, even allowed him to be the guest editor of an issue of the *Transatlantic Review* at a time when Hemingway was still a relatively unknown writer.

And later on in his life, in bad return, when Hemingway wrote his famous memoir, *A Moveable Feast*, his story of Paris in the '20s, he describes Ford joining him at his favorite table at the café in which he, Hemingway, preferred to write, and annoying him by showing up; describes him as a breathless old walrus. Well, he had been gassed in the First World War, and he always had trouble breathing, which was the reason that he was gasping the way he did. And Hemingway's memoir contains this contemptible line; he says, "I tasted my drink to see if [Ford's] coming had fouled it." Bad on you, Hemingway.

Let's turn to the novel, probably Ford's masterpiece, *The Good Soldier*. The most central and important fact about the novel, the issue that creates the most difficulty but is also the source of the book's importance and genius, has to do with his unreliable narrator and with the problem of irony that is connected to the fact that the narrator is unreliable and can't be believed.

The book was published in 1915 when Ford was 42, his partnership with Conrad nearly 15 years behind him. It was his first, and some would say it was his only, masterpiece, although I would include *Parade's End* among the masterpieces. It's a book that's so drenched in irony and indirection that it might almost be said to carry one principle of Modernism—that of the unreliable narrator—as far as it can go.

There is no voice of omniscience or authority in Ford's novel, and the voice of the narrator is peculiarly seductive—apparently totally guileless, candid, direct. He begins the novel with the famous line, "This is the saddest story I have ever heard."

Well, the problem of irony—of recognizing or of identifying the irony in the book—becomes a central and problematic aspect not only of *The Good Soldier*, but of Modernism itself. There are critics and readers even today, 90 years after the book's publication, who are still arguing over how to understand John Dowell, the narrator of *The Good Soldier*. One influential critic has actually argued that Ford's meaning is ultimately not ascertainable, because of the instability of his narrator's voice. I think this is a mistake, as I'll try to show.

But if we listen to that voice, what we hear is something strange and something disturbing. It's a voice that's full of false dichotomies, full of contradictions. It's full of confessions of ignorance, of helplessness. It offers us incessant rhetorical questions. At one point in the novel, for example, very early in the book, the narrator describes his relationship with the three central characters of the novel, and I suppose it's worth making clear what that basic relation is before I proceed.

The novel describes, essentially, a friendship between two couples: John Dowell and his wife, Florence, and an English couple, Edward and Leonora Ashburnham. Dowell explains in the course of the novel—he himself is an American—how he married this American girl, moves to Europe with her because she'd always dreamed of traveling in Europe, and while they're in Europe they meet the Ashburnhams; and they're traveling, really, from health spa to health spa, because his wife Florence has a heart condition, or an alleged heart condition, and his job is to care for her. At the very moment when he proposes to his wife Florence, she says to him, I have this heart condition. You have to know this, and although I'm willing to marry you, you must realize that we can never have conjugal relations. And Ford's narrator, John Dowell, says, "I was ready enough." This should tip the reader off to the idea that there's something not totally reliable, not totally normal about the narrator. But there are many other hints and suggestions much earlier in the novel that tell us the same thing.

The Ashburnhams, on the surface, look like the most magnificently happy and elegant couple one could imagine. And for nine and a half years, the Dowell's and the Ashburnhams carry on what Dowell tells us is a perfect friendship, a kind of minuet. At the end of this time John Dowell discovers that the entire relation was a fraud, that for most of that time his dear friend Edward Ashburnham was carrying on a sexual affair with his wife Florence with the full knowledge of

Ashburnham's wife Leonora, and that he, Dowell, was the only one who was in the dark about this.

In addition he discovers, although he doesn't tell us this in the beginning of the book of course, that his wife Florence had had affairs with other men before she took up with Edward Ashburnham, and that Ashburnham himself had had a history of sexual liaison with women other than his wife before he took up with the woman that Dowell describes in *The Good Soldier* as "poor dear Florence."

Well, that's the basic situation of the novel. And in the very beginning of the book Dowell reveals to us that for nine years they had had a perfect relationship, and then after nine years he discovered the terrible truth—although the full details that I have given to you don't emerge until later in the book, and they emerge obliquely, and they emerge accretively; sometimes they almost seem to emerge without the narrator's full understanding, as if he lets them drop by accident, or certainly without premeditation.

The very opening pages of the novel, as I've already suggested, however, should provide the reader with a clear sense that this narrator can't be trusted. As I've said, his voice is full of contradictions, of confessions of ignorance, and of confessions of helplessness. At one point very early in the book he describes their life before he'd found the truth as a beautiful minuet, a perfect minuet. And then within the next sentence he says, no, no, it was not a minuet; it was "a prison full of screaming hysterics."

Here's another example, also from very early in the novel. "And yet again you have me," says our narrator. If poor dear Edward—I don't know. If poor dear Edward was a noble person, he says, who could have doubted it? And then he goes on to say, No, Edward was not a noble person; he was a horrible conniver and adulterer. And he says, If this was true, if I couldn't understand Edward himself, "there is nothing to guide us. And if everything is so nebulous about a matter so elementary as the morals of sex, what is there to guide us in the more subtle morality of all other personal contacts, associations, and activities? Or are we meant to act on impulse alone?" he asks. "It is all a darkness."

The interrogative is this tentative, passive man's deepest mode. How could I see this? he asks. How can one know this? How can I express this? Who could ever make sense of this? This is the sort of question he asks obsessively. The first 15 or 20 pages of this novel probably contain more question marks than any other 15 or 20 pages in the history of English literature.

And again, as I've suggested, he's constantly contradicting himself: On the one hand, he'll tell us that his friend is a noble, good soldier; in another moment he'll say that he's a sentimental fool who's ashamed of his carnal desires. In one moment he'll describe his wife as "poor dear Florence," a heart patient and a sufferer. In the next moment he'll describe her as a manipulative creature who he hopes will burn in hell.

Though Ford never speaks in his own voice in the novel, I believe that the attentive reader will realize relatively quickly that the voice of this narrator cannot be trusted, is full of contradictions, is too insistent in confessing its inadequacy and puzzlement.

Take one more example, this from later in the novel, from the beginning of part 4: "I have," John Dowell tells us, "I am aware, told this story in a very rambling way so that it may be difficult for anyone to find [his] path through what may be a sort of maze. I cannot help it."

Well, the key to the novel is here, in the particular nature of what we have to recognize as Ford's drama of the telling. There's a drama of the telling in this novel, and it's a drama of helplessness. It's a drama in which the narrator is constantly confessing his utter inadequacy, his blindness, his ignorance.

As in *Heart of Darkness,* there's a kind of separate narrative here that parallels and even dominates the "traditional" story. In this case it's an adultery fable about the "perfect" friendship between the Ashburnhams and the Dowells, but it's told in a disrupted, achronological fashion in which major events are disclosed reluctantly and obliquely.

The reader is given a continuous, difficult task to reconfigure the story's chronology, to compare his own reconstruction of the events to Dowell's own judgments and opinions. This is an immensely more strenuous aesthetic and moral experience in some respects than what's demanded from more traditional novels. And what we might recognize here is how deeply and powerfully the Modern novel—Modernist fiction—is committed to what might be called the autonomy of its fictional world.

In Modern fiction, readers are not only not given simple exposition, they are expected in a kind of deep way to collaborate in the making of meaning, to reconfigure the events they read about and in some degree to complete them. The great theorist of all these questions is the American novelist Henry James, who spoke repeatedly in his famous prefaces to his novels of what he called at one point "the muted majesty of authorship." There is a majesty in it, but for James it's muted. And James was the first systematic theorist of Modern fiction.

For James, the importance of using a center of consciousness through which the story was narrated was crucially important for the sense of believability and plausibility that a novel would offer. James very powerfully felt that the omniscient narrator of the 19th century, the kind of narrator we associate with George Eliot or with Tolstoy, was factitious in certain ways, betrayed the extent to which the novel was a made-up artifact; and James argued powerfully, theoretically and by example in his own novels, for another way of thinking about fiction.

Both Ford and Conrad went to school on James. James was the man whom Conrad addressed in his letters as *très cher maître*, and the Jamesian idea of the muted majesty of authorship, the Jamesian idea of a center of consciousness through which the story would be told, was adapted by both Conrad and Ford to special effect and to powerful effect. Conrad and Ford's impressionism is a version of that Jamesian theory.

In *The Good Soldier*, the narrator's tentative, puzzled, contradictory voice and the evasive, circling shape of his story both express, we come to see, not a desire to understand, but an evasion of understanding. The drama of the telling here is almost a form of pathology; a psychological stratagem, a desperate effort at self-justification by a damaged, morally obtuse man whose wife and friends have betrayed him to his face for nine years.

We come to realize after a while that Dowell is telling the story not to illuminate anything, not to communicate with the reader, but to justify his life, desperately trying to pretend that he should not have known, that he could not have known what really was going on in his life. And it becomes clearer and clearer to the reader, of course, that the drama of the telling is an act of psychological reinforcement, is an act almost of pathological self-deception.

One way to crystallize some of this and some of the differences between some of the writers I've mentioned is to say that if we thought about Conrad's Marlow and some other Conrad narrators, but especially Marlow, we could say that Conrad is committed to what could be called a rhetoric of uncertainty, in which his narrator is not mad or crazy or pathological, as Dowell is in *The Good Soldier*, but is aware of the limitations of language and the limitations of memory; and the uncertainty with which he approaches his task, the tentativeness with which he approaches his task, is a function of that.

We could say that in the case of John Dowell, what we have is a rhetoric of helplessness or a rhetoric of impotence; in which the narrator seems committed to a theory of experience in which nothing can be known, and in which his own helplessness and his own obliviousness is perfectly understandable because the world itself is totally inscrutable. So, we could say that Dowell is committed to a rhetoric of helplessness.

I think, as we'll see when we move on to D. H. Lawrence, we could say that Lawrence is committed to another kind of rhetoric, a rhetoric of incantation, in which he tries almost like an ancient shaman to summon forth from the depths of personality and from what lies submerged beneath the surface of manners, instincts and energies that are otherwise unknowable or unseeable.

And as you will see when we get to Joyce, we could describe Joyce's rhetoric as a rhetoric of inclusion or the rhetoric of the encyclopedia, so elaborate and extensive is Joyce's ambition to include as much as possible.

And finally, another writer that we'll be talking about a bit later, William Faulkner, we can say that his rhetoric is a rhetoric of conjecture or speculation; in which the same principle of an unreliable narrator and a drama of the telling is used in a way that emphasizes the conjectural or speculative nature of the material he's describing. And when we turn to *Absalom, Absalom*, we'll see that a variation on what Ford and Conrad are doing occurs in that novel as well, but with a much greater emphasis on this principle of conjecture.

In any case, the full import of Dowell's drama of self-deception, its implication for the other characters and the novel's larger meaning, will be my topic in the next lecture.

One way to summarize what I've been saying about these Modernist texts in general, and especially about *The Good Soldier*, is to remind you that I've suggested that these Modernist texts demand a particular kind of collaboration from the reader, a much more active and strenuous behavior on the reader's part than has been the case previously.

One might note that this is a true kind of interactivity, and a far more complex and morally challenging kind of interactivity or interaction with the text than what is on offer in the computer games that are so often described as more democratic and interactive than the allegedly dead-tree technology of the printed book.

Lecture Ten
The Good Soldier—Killed by Kindness

Scope:

One key to understanding Ford's *The Good Soldier* is through the rhetoric of helplessness: the way in which Dowell's narration pivots on his recurring confessions of inadequacy and helplessness. In this lecture, we explore this notion more fully, applying it to the very structure of *The Good Soldier*. The drama of the telling in the book reflects and enacts in its own sphere a kind of helplessness. By studying some of the book's central passages, we discover how all the characters delude themselves and, in the process, destroy each other.

Outline

I. Two passages in *The Good Soldier* help us see how the drama of the telling mirrors the concept of hopelessness within its very structure.

    A. In the first passage Dowell is referring to a maid who has stolen from his family when he says, "Who in this world knows anything of another heart—or of his own?" He means that if one cannot know the heart of even a simple maid, one cannot know more complex matters that touch on human experience.

    B. By Dowell's assertion, in a second passage, that he could never have known Ashburnham's evil intentions, he elevates a personal failing to the level of universal law. If he cannot understand something, the world is not understandable.

    C. Dowell's view differs from that of Conrad's Marlow, whose example says that experience can never be fully or perfectly understood or captured in words, but this does not make the world totally unreadable.

    D. Conrad's vision of the world resembles the restless lament Coleridge expresses in the preface to his great poem "Kubla Khan."

        1. At the poem's start, Coleridge quotes a line in the original Greek from Theocritus, which translates as: "I

will sing a sweeter song tomorrow." Coleridge confesses that his poem is unfinished because he was interrupted while writing it and then lost his way.
   2. The implication of the allusion is that the poet is never satisfied and knows the world is ultimately elusive. But the poet will keep singing.
   E. Conrad respects the complexity of experience, but he does not think that renders it indescribable. Dowell's position, however, is that the world is unknowable. Why even bother, then, to try to say anything about it at all?

II. Another passage even more clearly dramatizes a similar aspect of the rhetoric of helplessness.
   A. A much more central incident in *The Good Soldier* is an account of the Dowells' and Ashburnhams' visit to the Castle of M.
   B. Dowell retells the visit several times, which in itself is characteristic of a certain kind of Modernist structure. In the first telling, readers get only part of the story because Dowell digresses.
   C. Dowell becomes aware of "something treacherous … in the day" but is helpless to act on this awareness.
   D. Leonora Ashburnham hurriedly leaves the castle, upset, and Dowell follows her. When they are alone, Leonora says to Dowell: "Don't you see what's going on?" She knows what Dowell will not acknowledge, that Florence is about to have an affair with her husband.
   E. At this point in the novel, we do not know that Ashburnham himself is a philanderer. Leonora is aware and, in some sense, has been complicit in her husband's affairs in order to maintain the façade of a happy marriage.
   F. One of the most crucial elements of the scene is that Dowell is reluctant to know the truth, preferring to be kept in the dark. Again and again in the novel, Dowell enacts that impulse toward a willed self-deception. The structure reflects this unwillingness because it takes multiple readings to clarify the extent of Dowell's self-deception.

III. Another important aspect of *The Good Soldier*, especially the scene at the castle, is how meaning is altered on second reading, a key feature of Modern fiction.

- **A.** As we said in Lecture Nine, readers of Modernist fiction are forced to engage in a much more strenuous interaction with the text and to engage in recreating the text, especially in novels with unreliable narrators.
- **B.** Readers must almost reorganize the material to create an accurate chronology because so many of these stories are told in an achronological fashion. They also have to see through the blindness or self-deception of the narrator's voice.
- **C.** On a second reading of the castle scene, we see more clearly Dowell's willful blindness and the history of pain behind Leonora's outburst. Thus, a second reading becomes a profoundly different experience from a first reading. It is as if there are two books to be read.
- **D.** As Ford had suggested, Modern fiction makes the reader work hard to understand meaning because life is that way; it does not explain itself to us. Our knowledge of real people and the world is an accretive process that never ends.
- **E.** On the other hand, Dowell's blindness in the dozens of moments when he is on the verge of truth but backs away may be natural or understandable for a passive, asexual person unable to bear the knowledge of horrible truths.

IV. Florence, Leonora, and Edward all enact their own miserable dramas of self-deception, paralleling and repeating Dowell's self-justifying narration, and, to some degree, all are killed by kindness.

- **A.** Dowell's marriage to Florence is fraught from the beginning. On the day he proposes to her, Florence says that they could not have sexual relations, and Dowell seemed to accept this condition.
- **B.** On their voyage to Europe, Florence has an affair with a young man and reveals that she has had other affairs as well. Florence, an American who longs to have the cultural authority of Europe, possesses a kind of American insecurity that Ford was fond of mocking.

C. She wants to appear more cultured and learned than she is. She even believes that the Ashburnhams' estate once belonged to her own family, but she will never get to see it. Ironically, her husband ends up possessing the estate, where he lives out his life playing nursemaid to another young woman, Nancy.

D. Florence is more interested in control than sex, which plays out cruelly when Leonora confronts her about her affair with Leonora's husband. Florence tells Leonora that she would leave Edward immediately if it would bring the two of them together.

E. Because she continues to forgive Edward for his series of affairs, Leonora demonstrates an even deeper form of self-deceptive, pathological behavior.

F. At no point does Leonora confront the possibility that Edward's dissatisfaction in their marriage might have anything to do with her. In fact, she elevates to the level of universal law Edward's inability to control his passions, deciding that all men are simply rutting animals.

G. She even tries to force Nancy—a surrogate daughter—to have an affair with Edward because she resents his last delusions of sexual decency and his failure to find her lovable.

H. So vicious is Leonora's forgiveness of what she pretends is her husband's animal essence that she drives Edward to suicide and, in a sense, destroys Nancy by such "kindness."

I. Edward deludes himself by believing that he is an English gentleman, a protector and chivalric knight in his relations with women. He is alienated from his own sexuality. Indeed, all the characters in this terribly revealing novel wreak horrific harm on one another out of an inability to understand or come to terms with their own sexual natures.

J. Ford's experimental impressionism and his essential material—balked sexuality and cruel intimacy—are truly innovative. Ford, and as we shall see, Lawrence and Joyce, colonized for serious literature the subject of sexuality and the physical life.

Supplementary Reading:

Ford, *A Man Could Stand Up*, *Last Post* (final vols. of *Parade's End*).

Friedman, "Point of View in Fiction: The Development of a Critical Concept," in Scholes, ed., *Approaches to the Novel*, pp. 113–142.

Questions to Consider:
1. How do Leonora's religious beliefs shape her response to her husband's infidelities?
2. What does the novel show or imply about the psychological mechanism of self-deception? Why do the characters deceive themselves about their own behavior and that of others?

Lecture Ten—Transcript
The Good Soldier—Killed by Kindness

I suggested in the last lecture that one key to *The Good Soldier* is what I called the rhetoric of helplessness, the way in which the tone of Dowell's narration depends upon a language which is constantly confessing its inadequacy and its helplessness. And I want to develop that notion a bit further and apply it to the very structure of *The Good Soldier*, to the way in which what I've called the drama of the telling in the book reflects and enacts in its own sphere a kind of helplessness.

There are two passages in the novel that I want to focus on that will help us see this as well as other matters. The first is a passage that comes fairly deep into the novel, and it's such a minor moment that many readers might pass over it as if it were trivial, unconnected to the rest of the book, or at least of minor importance. But, in fact, I think it's of critical importance. It's this passage. Dowell is speaking, as always, to the reader, and speaking in his usual interrogative and helpless way. He says:

> Who in this world can ever give anyone a character? Who in this world knows anything of another heart—or of his own? … That, for instance, was the way with Florence's maid in Paris. We used to trust that girl with blank checks for the payment of the tradesmen. For quite a time she was so trusted by us. Then, suddenly, she stole a ring. We should not have believed her capable of it; she would not have believed herself capable of it. It was nothing in her character.

And then Dowell goes on to say, essentially, if one can know so little about something so simple as the character of one's maid, how can one know anything about the deeper and more complex matters that touch on human experience? So, too, says Dowell, with Edward Ashburnham. How could I have possibly known that Edward Ashburnham, my close friend, my dear friend, who ate across the table from me virtually every night and who shared intimacies with me was at the same time carrying on a torrid and lurid sexual affair with my wife? How was that possible? And if that was true with Edward Ashburnham, so too with everything in the world.

What's going on in this passage is particularly interesting because it seems to me that one way to understand what Dowell is doing is to suggest that he is elevating a personal failing to the level of universal law. He's saying, I didn't understand something, ergo the world is not understandable. I missed certain clues. I was unable to figure certain things out; therefore the world, the universe is inscrutable.

One of the deep ways in which *The Good Soldier* has been misread by many contemporary readers has to do with this principle, because there are certain Modernist texts—we might think of the works of Samuel Beckett as an embodiment of these principles—that move toward a kind of nihilism, which suggests, in fact, that the world *is* an inscrutable, nihilistically oblivious place, a place that cannot be understood or cannot be interpreted, that the world is uninterpretable in some respects.

Dowell does say that, but that doesn't mean that Ford is saying it. And if we watch closely the structure of the novel and the shape of the novel, what we come to understand, in fact, is that this is Dowell's problem, that this is Dowell's failing, and that this principle of elevating a personal failing to the level of a universal law is also a behavior, a psychological trait of the other characters in the novel as well. I'll return to this matter in a while.

It's important to recognize how radically different this form of limitation, this way of confessing inadequacy is from, say, Conrad's Marlow's way of insisting on the world's complexity and the inexpressibility of experience. Marlow does say that experience is ultimately uncapturable in a perfect sense, ultimately can't be fully captured in language, and maybe can't be fully understood because our cognitive faculties are limited in certain ways. But Marlow does not imply that the world is totally unreadable. Something important, something complex, something rich is finally said. Something is finally known, even though what is finally known is imperfect and inadequate, and incomplete.

One way to crystallize this in terms of how Conrad looks at the world is to suggest that his vision of the world is something like that which is implied in the wonderful preface to Coleridge's great poem, "Kubla Khan." You remember I suggested that Conrad's *Heart of Darkness* had many things in common with Coleridge's "The Rime of the Ancient Mariner," and now I'm suggesting that the other great Coleridge poem, and maybe an even greater poem than "The Rime of the Ancient Mariner," "Kubla Khan," has all kinds of resonance, all kinds of connections to Modern fiction, and especially to Conrad's practice.

In the very beginning of that poem Coleridge quotes a line from Theocritus, the ancient pastoral poet. He quotes him in the original Greek, and it translates as something like this: "I will sing a sweeter song tomorrow." And this line follows a confession on Coleridge's part that "Kubla Khan" is an unfinished poem. He describes how he was interrupted in the writing of the poem by a visitor on business from Porlock, and then when the visitor left, the poem had gone out of his mind. And then, as if to say, well my poem is unfinished, my poem is incomplete—a shocking thing to be suggested about one of the great poems in the English language—but the beautiful thing, the crucial thing is that Coleridge also then appends the line, "I will sing a sweeter song tomorrow." I'll try again tomorrow.

The notion is not only that the world is not completely unknowable, but that one can try again to express its meanings. One can try again to capture its significance. This is characteristic, I think, of Conrad's attitude toward the world. He has a desperate respect for the inexpressible complexity of experience, but he doesn't think because experience is ultimately in perfect form uncapturable, that it can't be described, and that it is not worth describing.

Dowell's position—it is not Ford's—Dowell's position is that the world is unknowable, and the implication of that is why even bother to try to say it at all? Why even bother to try to understand the world or to describe it? So, Dowell's position is radically different from Marlow's, radically different from the romantic tradition which lies behind Modernism, and on which Modernism draws.

Dowell's helplessness in life, his blindness and passivity in the face of Florence's adulteries and the Ashburnhams' cankered marriage—these are mirrored, repeated, reenacted in the structure of his novel, in the drama of its making. And, as we'll see, this terrible impulse to understand one's own failings and weaknesses as part of the world's inscrutable systems is repeated yet again in the acts of the other characters in this terrible story of doom and destructive self-deception.

Let me turn to a second passage that dramatizes another aspect of this element of the novel even more clearly. This is the much more central incident in the novel, in which Dowell describes a journey that he made with the Ashburnhams and his wife Florence, a kind of outing to the Castle of M. The Castle of M has become a museum, and it's a museum of Protestantism. It's a place in which the history of Protestantism is registered, a place that Martin Luther once visited, and it exists now, in the now of the novel, as a kind of museum to these historic events.

The visit to the Castle of M is an event to which this novel returns several times in the course of its telling. And this is itself characteristic of a certain kind of Modernist structure. It's most reminiscent, I think, or most resembles especially the characteristic strategies of William Faulkner, and we'll talk about how Faulkner adapts these strategies very fully in *Absalom, Absalom*, as well as in some of his other novels.

In the case of *The Good Soldier*, what happens is that Dowell talks about the visit to the Castle of M at least three separate times in the novel, alludes to it, refers to it again; and, of course, the implication is that the first time he talks about it he tells a part of the story. Then he meanders around or digresses, or loses track of where he's going, and returns to it again later, and returns to it again later. Well, this principle of circularity, of returning again and again to the same modal points becomes a central feature of the structure of *Absalom, Absalom*, and it is in a certain deep way a central feature of the structure of *The Good Soldier* as well.

Here is part of the first moment in the novel where the visit to the Castle of M is described. Ford begins by describing the lovely train ride they took across the German countryside, and how restful it was to have Florence sitting happily on the train, chatting with Leonora and Edward so he, Dowell, in his role of nursemaid, in his role of caretaker for his wife who has allegedly the bad heart doesn't need his attentions, and how he was able to relax and look out the window and enjoy himself.

And then he describes their approach to the Castle of M and their entering the rooms, and he describes how alive and vital Florence is in this experience, how she sort of dominates the other three people, who are with her, in her characteristic way.

Let me read a bit:

> And Florence became positively electric, [Ford writes, or rather, Dowell tells us.] She told the tired, bored custodian

what shutters to open; so that the bright sunlight streamed in palpable shafts into the dim old chamber. She explained that this was Luther's bedroom and that just where the sunlight fell had stood his bed. As a matter of fact, [Dowell says—this is characteristic of him, to undercut his wife without acknowledging that he's undercutting her, as if he all through the novel really has deep resentments against her, but he doesn't allow his rhetoric to show it immediately.] As a matter of fact, I believe that she was wrong and that Luther had only stopped, as it were, for lunch, in order to evade pursuit. But, no doubt, it would have been his bedroom if he had been persuaded to stop the night. ...

"And there," [Florence] exclaimed with an accent of gaiety, of triumph, and of audacity. She was pointing at a piece of paper, like the half-sheet of a letter ... "There it is," [she says,] "the Protest." And then, as we all properly stage-managed our bewilderment, she continued: "Don't you know that is why we were all called Protestants? That is the pencil draft of the Protest they drew up. You can see the signatures of ... Luther, and Martin Bucer, and Zwingli, and Ludwig the Courageous."

And then follows another paragraph in which Dowell undercuts her historical knowledge, and then he says:

She continued, looking up into Captain Ashburnham's eyes, [and then she says]: "It's because of that piece of paper that you're honest, sober, industrious, provident, and clean-lived. If it weren't for that piece of paper you'd be like the Irish or the Italians or the Poles, but particularly the Irish ..."

And she laid one finger upon Captain Ashburnham's wrist.

Then Dowell says, "I was aware of something treacherous, something frightful, something evil in the day. I can't define it," here his rhetoric of helplessness, always present—I can't define it. "I can't find a simile for it." And then, characteristically, he does find a simile for it.

I can't find a simile for it. It wasn't as if a snake had looked out of a hole. No, it was as if my heart had missed a beat. It

> was as if we were all going to run out and cry; all four of us in separate directions, averting our heads.

Remember, this is at the very beginning of the relationship between the Ashburnhams and the Dowells.

> In Ashburnham's face I know there was absolute panic. I was horribly frightened and then I discovered that the pain in my left wrist was caused by Leonora's clutching it:
>
> "I can't stand this," she said with a most extraordinary passion; "I must get out of this."

And then Dowell says:

> I was horribly frightened. It came to me for a moment, though I hadn't time to think of it, that she must be a madly jealous woman—jealous of Florence and Captain Ashburnham, of all people in the world! And it was a panic in which we fled!

He then follows Leonora out of the room, out of the museum, to another place, and he's alone with Leonora Ashburnham, and she says to him:

> "Don't you see? ... Don't you see what's going on?" The panic stopped my heart. I muttered, I stuttered—I don't know how I got the words out:
>
> "No! What's the matter? Whatever's the matter?"
>
> She looked me straight in the eyes; and for a moment I had the feeling that those two blue discs were immense, were overwhelming, were like a wall of blue that shut me off from the rest of the world.

And the passage goes on for a while, describing in some detail how complicated his feelings were. And Leonora then responds to him, and then the scene ends; the passage ends. This is very early in the novel, in part 1 of the novel.

> She ran her hand [Leonora] with a singular clawing motion upwards over her forehead. Her eyes were enormously distended; her face was exactly that of a person looking into the pit of hell and seeing horrors there. And then suddenly she stopped. She was, most amazingly, just Mrs.

Ashburnham again. Her face was perfectly clear, sharp and defined ... Her nostrils twitched with a sort of contempt. She appeared to look with interest at a gypsy caravan that was coming over the little bridge [toward] us.

"Don't you know," she said, in her clear hard voice, "don't you know that I'm an Irish Catholic?"

Now, what's happening in this scene is very remarkable. This may in fact be Leonora's most generous moment in the entire novel. What has happened, of course, is that Leonora sees in this moment that Florence is about to annex Edward Ashburnham, and she sees that Edward is succumbing to Florence's charms. And she rushes out of the room in a horror and a rage.

At this early point in the novel the reader doesn't know what we learn later, that Ashburnham himself has a whole history of affairs with other women that Leonora, herself, has known about and in some sense has been complicit in because she's tried to protect her name and her marriage from scandal. She's much more interested, in some degree, in maintaining the conventional façade of a happy marriage than she is in actually having a happy marriage.

As we learn later in the novel, as Dowell reveals things obliquely and in parentheses, almost reluctantly, Leonora has known about Edward's adulterous proclivities for a very long time. So, when we encounter this scene for the first time, we readers have no idea of any of this history, and, of course, Dowell has no idea of it either. However, one can see that Dowell can certainly recognize that Leonora is horrifically upset, and Leonora is on the verge of saying to Dowell, Don't you see what's going on? Your wife is about to have an affair with my husband.

And as she continues and she watches Dowell, and she sees how Dowell's responding, she suddenly realizes that Dowell doesn't want to know, that Dowell wants to be kept in ignorance. And we can see Dowell moving toward a recognition of what's happening and then backing away from it, not wanting to know. So when her nostrils twitch with a sort of contempt, the contempt is for Dowell, and for Dowell's desire to remain in the dark.

And when she says "Don't you know that I'm a Irish Catholic?" what she's doing is giving another explanation for why she was so offended, because, of course, Florence's remarks, as you will recall, were about Roman Catholics, were about Catholics and about being Irish. So she uses the fact of her religion as another way of explaining why she's so upset, when of course what she's really upset about is what she decides not to reveal to Dowell, who prefers to remain in the dark.

A number of things about this scene are crucial, central to the novel. One of the most central things I've already suggested is that we see Dowell desperate to be deceived, desirous of deception, reluctant to know the truth, preferring to be kept in the dark. And we see the extent to which he is almost brought to a recognition of what's really the case and backing away from it. Again and again in the novel we will see this happening, not only in what might be called the traditional story, the events that occurred before the narration has begun, but we will see Dowell enacting that kind of impulse toward a willed self-deception in the structure of his storytelling itself, in what I've called the drama of the telling, or the drama of the narration.

One of the most important things to say about this astonishing scene is how its meaning alters on second reading. This is a key feature of Modern fiction. A second reading of many books in Modern fiction shows you almost an entirely new book because, as I suggested in the last lecture, the reader of a Modernist text is forced to engage in a much more strenuous kind of interaction with the text, is forced in some sense to engage in a form of recreation of the text—especially when the novels are narrated by unreliable narrators—is forced in a certain sense to reorganize the material, to create an accurate chronology, because so many of these stories are told in an achronological fashion. And the reader is also, of course, often forced to read through the blindness or the self-deception or the unreliability of the narrator voice, so that that means by the time one finishes the novel and has recreated all the events, has got things straight, what we realize on second reading of this scene is all that history that I've mentioned a moment ago.

The second time we read the scene and we hear Leonora say, "I can't stand this," or we hear Dowell saying, Ah, I thought there was something treacherous in the air, suspecting Florence and Captain Ashburnham, of all people, what we see is not only his willful blindness, but the tremendous history of pain and unhappiness that lies behind Leonora's outburst at this moment. So, the second reading of this scene, just like the second reading of the novel, becomes a profoundly different experience from a first reading; and it's one of the great features, one of the signature attributes of Modern fiction that this is the case.

Our awareness of the ironic implications of the story only becomes clear after a successful careful first reading, and it makes a second reading often much more powerful and resonant for us. It's as if there are two books to be read. The second time through one is sort of clued in, one has done one's homework, and one is able to savor ironies and complexities that one is oblivious to on first reading—inevitably oblivious to.

Why does Modern fiction make the reader work so hard, collaborate, really, in the making of meaning? As Ford had suggested, and as I mentioned in the previous lecture, because life is that way. Life doesn't explain itself to us. Strangers don't introduce themselves to us with expository paragraphs summarizing their lives, telling us what went on to this point of time in their experience. We learn the world as we learn Modern fiction: slowly, after being puzzled, putting fragments together, positing interpretations that we then alter as new information emerges. Our knowledge of real people, of the world, is accretive; it's a process; it's never finished; as is true of these books, and the best of these books as well.

But I don't want to go too far in emphasizing Dowell's blindness and Dowell's folly, because there's another way to think about it, maybe a more humane or generous way to think about it. There's one sense in which Dowell is a kind of reluctant truth-teller. Even in the scene I've just talked to you about, we can see Dowell moving toward a recognition of the truth and then pulling back from it.

And if you read *The Good Soldier* carefully and attentively, you will see that there are dozens of such moments in the text, in which Dowell moves closer and closer to the very verge of understanding, and then backs away. I think one way to understand this is to say that the knowledge that he would have to bear, that he would have to confront, is so burdensome, so terrible, so awful—that his whole life is meaningless, that his life has been a fraud, that he has been played upon and betrayed by the people who are closest in his life—that those facts are so terrible, so awkward, so hard for him to accept, that it's humanly natural for him to back away from them.

And I think that the novel makes this argument—again obliquely, as it makes all its arguments—in another way, because of the way it treats the other primary characters in the book. And I want to conclude this lecture by talking a bit about those final characters, about Florence and Leonora and Edward, who all might be said in some degree to be killed by kindness.

I've already spoken about the plot, and it's worth reminding you that in some degree all the other characters in the novel enact their own miserable dramas of self-deception and self-damage. Let's talk a bit about Florence's marriage to Dowell, quickly. We learn, as I think I mentioned earlier, that Dowell's marriage to Florence is fraught from the beginning. On the very day that he proposed to her, Florence told him that they could not have sexual relations, and he said, "I was ready enough."

It turns out, as is ultimately revealed to us in Dowell's narration, that on the voyage to Europe, aboard ship, Florence was carrying on an affair with someone else, with a young man who carried their baggage, and that she had had other affairs as well. What we sense about Florence over and over again, as more and more details accrue, is that she has a kind of American insecurity, that she belongs to a particular kind of American that Ford was very fond of mocking and making fun of, an American who longs to have the cultural authority of Europe. And so her bossiness in the scene at the Castle of M is an emblem of her character. She wants to seem as if she knows much more than other people, as if she is learned and cultured in a way that, of course, she's not.

One of the great ironies of the novel is that we learn that Florence actually has the delusion that the ancestral estate owned by the Ashburnhams was originally in her family, and she has dreamed of getting back to her ancestral estate. A further irony is that, because of her lie about her heart condition, once they get to Europe they are advised by doctors—or at least Florence manipulates things such that they are advised by doctors—that the crossing of the English Channel would now be too dangerous for her bad heart. So, she can never even make it to England to visit what she imagines to be her ancestral home.

The greatest irony of all, of course, is that after all the tragedies and suicides and deaths with which the novel concludes, we are left at the end of the novel with Dowell himself in possession of the

Ashburnham estate, again returned to the condition of nursemaid, now nursing another woman, a young woman named Nancy, who has gone mad because of her connection to the Ashburnham inferno.

So Florence's marriage to Dowell reveals certain things about her that are then acted out elsewhere in the text. She's not really interested so much in sex as in control. That's her deepest need. And there's a terrible, memorable scene late in the book where Leonora comes to her—Leonora has always been aware of Edward's affairs—and confronts her, and Florence says to Leonora, Oh, Leonora, I would leave Edward immediately if it would bring you two together. And Leonora says something like, You come from my husband's bed to tell me such a thing? And of course what we feel is that Florence gets more pleasure from being able to manipulate Leonora in this way than she does from her affair with Ashburnham itself.

We can see an even deeper form of self-deception and pathological behavior in Leonora, because Leonora keeps forgiving Edward for his series of affairs. She intervenes in some ways at a certain point in the history of their marriage to bring him into contact with a respectable younger woman whom she knows Edward is in love with, because she prefers that Edward have contact with respectable women rather than with prostitutes, or with unsavory women of the lower social orders. It's as if she comes to pimp for her own husband.

At no point in the book does Leonora ever confront the possibility that Edward's dissatisfaction in their marriage might even remotely have anything to do with her. And, in fact, she elevates to the level of universal law Edward's inability to control his passions. She decides that all men—and she has a special priestly advisor to help her to believe this—she decides that all men are simply rutting animals.

The worst and most terrible thing that she does in some respects is that she forgives Edward. She goes to Edward and, essentially, she says to Edward, "I know you can't help yourself, so carry on." And the effect of this is to destroy Edward's own sense of himself, because now let's make a transition to Edward.

Edward has the delusion that he is a kind of throwback to the 18th century, that he is a kind of knightly English gentleman. This is part of his family tradition. He is a member of the squirearchy, of the landed gentry of England. And what is dearest to Edward's heart is the idea that he is a protector and a chivalric knight in his relations to women. Every time Edward feels sexual desire he tries to translate it into a chivalric feeling. He, himself, is alienated in some sense from his own sexuality.

So, Edward and Leonora themselves are also the victims of deep, deep forms of self-deception. Leonora's most brutal act is forcing her ward, Nancy, who was, in effect, their surrogate child, to offer herself sexually to Edward. This destroys Nancy's innocence. Nancy reveres both of them, and at a certain point Leonora goes to her and says, Edward can't help himself. Give yourself to him. He needs you. And this destroys her innocence. So vicious is her forgiveness, she kills Edward and in a sense destroys Nancy by kindness.

So, all the characters in this terribly revealing novel wreak horrific harm on one another—Dowell retroactively, in his narrative—out of an inability to understand or come to terms with their own sexual natures.

We might want to end, then, by emphasizing how innovative Ford's subject matter really is; not just his experimental Impressionism but his essential material, his story of balked sexuality and cruel intimacy. Ford, Lawrence, and Joyce especially, colonized for serious literature the subject of sexuality and the physical life. This side of Modernism is often overlooked, receives less emphasis than the often off-putting technical and stylistic difficulty for which Modernism is famous; but the great Modern novelists were the ones who opened the novel to the wonders and miseries of our sexual lives.

Lecture Eleven
Lawrence (and Joyce)—Sex in Modern Fiction

Please be advised that parts of this lecture contain some explicit discussion of sexual matters and may be unsuitable for children.

Scope:

This lecture places D.H. Lawrence's treatment of the themes of alienation and sexuality in a context that includes Ford Madox Ford as well as Marx and Freud. After a brief sketch of his life, we turn to an account of Lawrence's major fiction, emphasizing his own conflicted sexuality and the candor of his writings about sexual thoughts and behavior. We conclude with a close and perhaps still shocking look at a moment of sexual intimacy from *Lady Chatterley's Lover*. Is the tenderness Lawrence intended us to see there still present for contemporary readers? Whatever the answer, Lawrence and Joyce are comrades in a defining Modernist project—the exploration of our sexuality.

Outline

I. Estrangement or alienation is a major theme in Modern literature.
   A. We can define alienation is the disconnection between social conventions and moral norms on the one hand, and human needs and appetites on the other.
   B. An awareness of alienation probably originates in the 19th century and Romanticism. The theme takes on compelling power in Modernist literature.
   C. As we saw in Lecture Ten, the characters in *The Good Soldier* are victimized by the destructive irrelevance of inherited notions of sexual and personal conduct.
      1. The fault, the book implies, is the failure of religious and social institutions to provide a framework for real human nature.
      2. The gap between our actual selves and the moral and social systems of pre-World War I Europe is at the heart of this novel's darkness. Ford emphasizes these

connections by using August 4 (though in different years) as the date for most major turnings in the story—deaths, suicides, and significant anniversaries. This is the day that Britain declared war on Germany.

- **D.** Alienation was also a crucial theme for such thinkers as Marx and Freud.
    1. For Marx, alienation is a fundamental theoretical category. His writings attempt to define the nature of our estrangement from work, nature, ourselves, and others.
    2. For Freud, alienation means we are cut off from our deepest nature, our inner self. Human beings are driven by needs and appetites to which they are oblivious or that they cannot acknowledge.

II. Although only 45 when he died, D. H. Lawrence (1885–1930) was immensely prolific, and both his life and his work are still controversial today. Some have attacked his notions of sexuality, the relationship between human beings, and the social order, while others have accused him of fascism.

- **A.** Lawrence's father was a barely literate coal miner, and his mother a schoolteacher with literary and cultural interests. His parents' unhappy marriage inspired some of his most interesting work, much of which includes working-class themes.
- **B.** Lawrence wrote 12 novels and one longer novella, *The Man Who Died* (1931), which is sometimes considered a thirteenth novel. The latter is a retelling of the Christ story, offensive to many because it reimagines the Resurrection as having a carnal, rather than a spiritual, aspect.
- **C.** Lawrence also wrote some 60 short stories and nearly 800 poems. He was the author of a series of illuminating travel books which some believe represent his greatest writing.
- **D.** At the end of his life, Lawrence took up oil painting, creating Cézanne-like images with allegorical characters and an emphasis on human carnality.
- **E.** His most significant novels are *Sons and Lovers* (1913), *The Rainbow* (1915), and *Women in Love* (1920). The first is largely autobiographical; the other two were banned for their open description of homosexual relations.

III. *Sons and Lovers* is one of Lawrence's most remarkable achievements and highly recommended for those readers who want a sense of his work as a novelist.

  A. *Sons and Lovers* dramatizes the story of Lawrence's mother's death from cancer, as well as his relationships with other women. The novel explores sexuality and psychological dependency through its central character, Paul Morel.

  B. Morel's intense, psychologically intimate, and dependent relationship with his mother reflects Lawrence's own relationship with his mother and has disturbing undercurrents.

  C. Morel is unable to commit to a woman his age with whom he has an extended affair. The implication is that his dependence on his mother is so deep that he cannot completely connect with other women.

  D. Later, he has an affair with an older married woman who, to a degree, represents his mother. Although the book is not openly perverse, it has strong Oedipal elements.

  E. *Sons and Lovers* articulates a central Lawrentian theme: the idea that selfhood is an unfinished project. One is always in the process of becoming; the effort at serious living is fraught with difficulties and moral peril.

IV. *Women in Love* is another of Lawrence's most notable and influential works.

  A. *Women in Love* is marked by postwar bitterness and contains homoeroticism beyond that depicted in *The Rainbow*.

  B. Each chapter in the book is a separate, symbolic vignette. The story revolves around two sisters, Ursula and Gudrun, whose family also appears in *The Rainbow*. In *Women in Love*, Ursula's lover is Birkin (a character associated with Lawrence himself), and Gudrun's lover is Gerald.

  C. The novel dramatizes homoerotic tensions between Birkin and Gerald, especially clear in a scene in which the two naked men wrestle together.

  D. At the end, Ursula and Birkin are confident of their committed love for each other. But in the final lines, Birkin confesses his desire for a perfect relationship with another man. Birkin says (in paraphrase), "I can't but I wish I could; I don't want to believe that I couldn't." Ursula dismisses this as ridiculous. Birkin's actual last line is, "I don't believe that."

E. *Women in Love* ends, then, on an uncertain, dissatisfied note, in which we feel Lawrence himself struggling with his own homosexual tendencies.

F. Another scene in *Women in Love* captures an aspect of Lawrence's writing that can help us see more deeply into a related aspect of his work. In this scene, Ursula witnesses a country wedding as the party assembles outside the church.

  1. When the groom arrives, Lawrence describes almost in animal terms how the groom runs toward his bride: "leaping the steps and swinging past her father, his supple haunches working like those of a hound that bears down on a quarry."

  2. This animalistic imagery pushes the scene beyond the social conventions it depicts to the instinctual energies underlying the act of marriage itself.

V. In today's openly sexual culture, it is easy to underestimate one fundamental part of Modernist fiction's legacy—its pioneering insistence on speaking openly about sexuality. Such writers as Joyce and Lawrence, however, hold a place of honor in literary history for their exploration of our sexual lives.

A. Lawrence's *Lady Chatterley's Lover* was first published in 1928 and later banned; it was legalized again after trials in 1959 (America) and 1960 (Britain).

B. The story is a fable of social class in which Constance, the wife of a wheelchair-bound industrialist, has an affair with a gamekeeper on their estate.

C. The book's sexual candor and its transgression of class barriers were shocking in their day.

D. The original title for *Lady Chatterley's Lover* was *Tenderness*, and the book actually contains many tender moments, including one in which Mellors admires and touches Connie's nether parts. Despite his Yorkshire dialect and his perhaps comic bluntness, can we hear the wonder and the tenderness in his speech? (Lawrence knew the Yorkshire dialect intimately and wrote a number of poems in dialect, as well.)

E. Lawrence dramatizes himself in a way that sees through his own pomposity and didacticism. The power in his novels arises from the fact that his characters are often the ones who are subject to the most withering criticism and mockery.

F. Both Joyce and Lawrence are comrades in the great intellectual struggle to open literature to the full range of human experience and, in particular, to understand and dramatize our sexual natures.

Supplementary Reading:

Lawrence, *Sons and Lovers* and *Lady Chatterley's Lover*.

Questions to Consider:

1. What developments in the early 20th century might have led writers and thinkers to see estrangement or alienation as a special problem?

2. How is Lawrence's vision of alienation related to his attitudes toward sexuality?

Lecture Eleven—Transcript
Lawrence (and Joyce)—Sex in Modern Fiction

I left implicit in my account of *The Good Soldier* the links between the sexual confusion and misery of the primary characters and the larger social order, but Ford Madox Ford certainly intends his readers to perceive such a connection. His characters are victimized by the inadequacy, the destructive irrelevance of inherited notions of sexual and personal conduct.

The fault, the book implies, is both in religious and in social institutions; the failure of the institutions of class and of religion to provide frameworks or conceptual maps for real human nature, for the needs and appetites that human beings discover within themselves. The gap between our actual selves and the moral and social systems of pre-World War I Europe is at the heart of this novel's darkness.

Ford emphasizes these connections by a strange and, actually, in some ways implausible principle of coincidence. Most of the major turnings in the story, the deaths and suicides of some of the characters, the significant anniversaries, all occur on August 4th, though of course in different years. And Ford expected his readers to recognize that that date, August 4th, was the day on which Britain declared war against Germany, the start of World War I.

One name or label for this disconnect between social conventions and moral norms, on the one hand, and human needs and appetites on the other, is estrangement or alienation—one of the defining themes of Modern literature and art. It's a recurring intuition or affliction for the modern sensibility, this feeling or experience of alienation. Its origins go back deep into the 19th century and in the movement we call Romanticism; but the theme of alienation, the experience of estrangement takes on a particular compelling power in Modernist literature.

The condition of not belonging, the feeling of homelessness or of disconnection, is a crucial theme not only for our writers but, of course, for thinkers like Marx and Freud. For Marx, alienation is actually a fundamental theoretical category, and in his writings he is explicit and systematic in trying to define the nature of man's estrangements from his work, from nature, from himself, and from other human beings. Marx's central diagnosis, really, of the human plight under capitalism revolves around, centers on, the idea of alienation.

For Freud, too, this notion is crucial, although I don't know that Freud uses the term with exactly the same power. But Freud's account of human character involves a basic assumption of estrangement from one's deepest nature, from one's inner self. It's as if Freud's model of mind and model of selfhood involves an idea that conscience—the part of our minds that encourages us to moral distinctions and moral behavior—and consciousness itself—just awareness, consciousness—both consciousness and conscience are in Freud's account always oblivious to unconscious drives and yearnings. And his account of human nature often involves this idea of human beings being driven by needs and appetites and energies that they are oblivious to or cannot acknowledge.

This theme of alienation or estrangement is much more pronounced and central in European Modernism than in the English or American versions of Modernism with which I am familiar, because European Modernism is darker in many ways and more despairing. But there are, of course, central elements of this same almost universal theme in English and American writing as well.

One of the great prophets of alienation, one of the great dramatizers of alienation in a variety of complex forms is the writer I want to take up in the next two lectures, D. H. Lawrence. In one sense we might think of him as a poet and prophet of estrangement, at least among the English writers. He had a restless, disturbing, and prolific life.

He was born in 1885 and died in 1930, only 45 years old when he died, but he was immensely prolific, and his work is still to this day controversial, deeply controversial. There are feminist critics, for example, who have violently and angrily attacked his notions of sexuality. I think there's a good deal in their complaint, in fact. And there are others who have complained bitterly, perhaps also with justification, about the extreme forms that some of his ideas about the relationship between human beings and the social order take.

There are some people who have found in some of his later novels—they are sometimes called the leadership novels—a tendency toward fascism. I'm not positive that those tendencies are actually there, but the heart of Lawrence is the opposite of fascist, it seems to me. His legacy, in any case, is not settled. The value and importance of his work is still a matter of argument and intense controversy, as it was, as I've suggested, during his lifetime.

He was the son of a coal miner who was barely literate, and his mother was a schoolteacher who had literary and cultural interests; and that conflict between his mother and father became one of the central elements in his work. There are working-class themes in nearly all of his novels and stories, and at the heart of his very best stories and novels. And his parents' unhappy marriage provided, as I've suggested, other crucial fodder for his most interesting work.

He was the author of 12 or 13 novels. The confusion has to do with the question of whether or not certain longish stories that we might identify as novellas deserve the separate title of novel, because one of them at least was published by itself under a single title. It's a very interesting late, long, long story or novella called "The Man Who Died," and in many ways it's a characteristic Lawrentian tale, because it's a Christ fable, it's a Savior story, it is a retelling of the Christ story, but it imagines or re-imagines the Resurrection as having a carnal rather than a spiritual aspect. And there are many people who have found that late novella morally offensive because of its emphasis on the carnality of the resurrected Christ.

He was the author of some 60 short stories, of nearly 800 poems. The most admired of his poems are very intense evocations of animals and natural experiences. The title of his most well-known collection of poems is *Birds, Beasts and Flowers*. He was the author of a series of luminous travel books, and there are some readers and literary critics who believe that his greatest writing is in these wonderful travel accounts.

At the end of his life Lawrence also took up oil painting and painted very powerful, Cézanne-like images that have a kind of allegorical character. One of his paintings, for example, is called the *Holy Family*, and it shows ordinary people, in some degree, and the female in the *Holy Family* is bare-breasted. He loved in his paintings to emphasize human carnality, and some of his paintings were confiscated and caused great scandal because he represented pubic hair in the paintings.

His most significant novels, the ones for which he is most famous and the ones I think that are most important for any reader to examine are these: *Sons and Lovers*, an autobiographical novel published in 1913 and in some respects his most powerful novel (or at least many people feel that, and I do), and then a pair of novels published in 1915 and then in 1920 that are really linked. They were originally conceived as one single chronicle novel and they finally split into two parts, and their connection to each other is much less thoroughgoing and much less compelling, I think, than Lawrence may have originally intended.

The first of these is called *The Rainbow*. It was published in 1915, as I said, and it was banned in England because it contained homoerotic scenes of female relationships. It was one of the first novels in English to openly describe lesbian relations. The novel *Women in Love* came out in 1920, and I will say a few words about *Women in Love* in a moment.

But I'd to give you in this brief sketch of Lawrence's career at least a short account of *Sons and Lovers*, which remains, I think, one of his most remarkable achievements, and certainly the novel I would recommend to people, for those readers who wanted to begin to get a sense of Lawrence's work as a novelist.

The book is, as I mentioned, deeply autobiographical, and it actually dramatizes the story of his mother's death from cancer, and of the young Lawrence's relationships with other women. The novel is deeply about sexuality and psychological dependency. The central character in the novel, a very Lawrentian-type character—a young would-be artist named Paul Morel—has an intense, psychologically intimate, even dependent relationship with his mother, as Lawrence himself seems to have done. And the novel is therefore full of a kind of psychosexual undercurrent that's both disturbing and powerful.

This young man, Paul, conducts an extended affair with a young woman his own age who loves him deeply and is very attentive to him, and he has much in common with her, but he finds that he can't commit himself fully to her, and the implication is because his dependence on his mother, his connection to his mother, is so deep and so both empowering and disabling that he's unable to make a complete commitment to other women.

He then picks up in an affair with a married woman, and that relationship has all kinds of disturbing undercurrents, and it is almost as if this older woman is in some degree a kind of substitute for his mother. These pressures are in the book without it sort of treading over into any form of perversity, and without really explicit sexual descriptions of any sort, the sort of thing that Lawrence would become infamous for later in his career. But there are definitely sexual and Oedipal elements in the novel that are very powerful.

It is also a novel that articulates what might be said to be one of the central if not the central element in Lawrence's fiction and especially in his novels, which is the idea that selfhood itself is a project, a striving; that the self is never finished and is always in some process of becoming.

In the novel *Women in Love*, the Lawrentian protagonist—Birkin his name is, also a character patterned on Lawrence himself—speaks at one point of the effort. There's a line in the novel which goes something like this: Birkin felt that he was doomed to the effort at serious living. And the project of serious living, the effort at serious living, understood to be both a rigorously dangerous project and also one that's fraught with difficulties and fraught with moral peril, is at the heart of the emergent self that the young artist, Paul Morel, struggles to find in *Sons and Lovers*.

Women in Love, published in 1920, is full of a kind of post-war bitterness, and it contains more homoeroticism beyond what was present in *The Rainbow*, but this time the homoeroticism is primary male. There was a suppressed introductory chapter that was explicitly homosexual and that Lawrence himself seems to have suppressed.

Women in Love is a very remarkable book, still to this day. It's written in an immensely powerful series of symbolic scenes that have very short titles. The chapters almost are autonomous little vignettes or short stories, and the novel concerns essentially a juxtaposition of two couples, two sisters, Ursula and Gudrun, whose family is the family that both *The Rainbow* and *Women in Love* tell the chronicle of. Ursula and Gudrun each have a man in *Women in Love*. Ursula hooks up with Birkin, the Lawrentian character, the D. H. Lawrence stand-in, and Gudrun picks up with a man named Gerald, who is Birkin's best friend. And there are all kinds of homoerotic tensions and energies that are expressed in the novel concerning Birkin and Gerald, even one very disturbing and powerful scene in which Gerald and Birkin wrestle naked together on the floor of Birkin's cottage.

At the very end of the novel, after Ursula and Birkin have broken through to a full, complete relationship and are voyaging on into the future, absolutely confident of their love for each other, the very final lines of the novel are characteristic of the unfinished becoming quality that's characteristic of Lawrence's work. Ursula and Birkin are talking to each other, and Ursula says, Oh, I'm so happy that we're finally absolutely together, and we're beyond our quarrels, and Birkin says, Yes, I agree, but I still wish I could have a perfect relationship with another man. And Ursula says, That's perverse. That's ridiculous. You can't have such a thing. And the last line essentially says, Perhaps I can't, but I wish I could; I don't want to believe that I couldn't.

So the novel ends on a kind of uncertain note, and on a note, in a certain sense, of dissatisfaction, in which one feels also that Lawrence himself is still struggling with the homosexual tendencies that he felt were within him.

There's one scene in *Women in Love* that I want to call your attention to in this lecture that captures a certain aspect of Lawrence's practice as a novelist that can help us see more deeply into his work, I think. It's a scene that comes relatively early in the novel, and it's a scene that's witnessed by Ursula, not a scene in which she participates. She's watching, from a kind of hillside, a rural wedding. At the outside of a church a wedding party has assembled. The bridegroom is a bit late when the scene begins. And then this comes:

> The carriage rattled down the hill, and drew near. There was a shout from the people. The bride, who had just reached the top of the steps, turned round gaily to see what was the commotion. She saw a confusion among the people, a cab pulling up, and her lover dropping out of the carriage, dodging among the horses and into the crowd. "Tibs! Tibs!" she cried in sudden, mocking excitement.

And then Tibs arrives. It's her bridegroom. And Tibs is described almost in animal terms, "leaping the steps and swinging past her father, his supple haunches working like those of a hound that bears down on a quarry."

And there's much more of this, in which we see Lawrence turning the scene into almost a kind of a scene of animal activity in which he wants to in some sense press past, press beyond, the social conventions to the animal instinctual energies that lie beneath the social convention for which marriage presumably stands, or which marriage is supposed to channel or consummate.

This element in Lawrence is a crucial aspect in his work, and it also causes readers, I think, often to be disturbed or surprised by what he does because it's as if his characters behave in surprising or in many ways inexplicable ways. In some cases they seem unmotivated, and in the great story, "The Horse Dealer's Daughter," that we're going to consider in the next lecture, I'll try to develop these ideas more fully and show how they operate in a systematic way in one of Lawrence's greatest short stories.

What lies beneath the social conventions that Lawrence is trying to dramatize in that wedding scene, in which the groom turns into an animal in the hunt and in which the bride exultantly embraces the role of the hunted, and her own animal nature is dramatized as well—what lies beneath that, of course, is something of Lawrence's notion of the power of sexuality. And the sexual candor of Modern fiction is something I want to talk about a bit now.

It's a fundamental part of the legacy of the great Modern writers, and it's very easy to underestimate or even miss entirely today in our culture of X-rated celebrity and instant pornography by mouse-click. But Lawrence and Joyce especially would hold a place of honor in literary history for their part in one of the great Modernist projects, the exploration of our sexual lives, even if they had no other claim to be remembered. The great instance of this in Lawrence's work is *Lady Chatterley's Lover*, of course, a novel first published in 1928, legalized in trials 30 or more years later in New York City in 1959 and in London in 1960.

Lady Chatterley's Lover remains something of a scandal even today. It is a fable of social class in which an industrialist named Sir Clifford Chatterley, who is impotent and sits in a wheelchair for the entire novel, is juxtaposed against his gamekeeper, the man who takes care of the game and manages the forests on his estate, a man named Mellors. And, of course, as the novel unfolds, Constance Chatterley (Lady Chatterley) sees Mellors in the woods. She's sexually frustrated. Her sexual energies have been balked and thwarted through her life, and she begins what becomes an immensely intense affair with Mellors—still one of the scandalous books in Modern literature.

We might note parenthetically that Joyce had problems with the censors that parallel those of D. H. Lawrence and that Joyce's treatment of sexuality has an equal if not a greater candor. These are matters to which I expect to return in later lectures, when we get to Joyce himself.

When Constance Chatterley first spies Mellors in the woods, she's not herself even aware of an interest in him, and one of the shocking things about the novel, of course, is not only that it is so sexually candid, but it also dramatizes another kind of transgression, which is the transgression across social class. Here is an aristocratic woman making love with, having an intense relationship with a member of the lowest social order, a man who's virtually illiterate.

And there are many moments in the novel, especially moments of tenderness, where Mellors falls into what I guess we might call a kind of demoted dialect. He begins to speak in a kind of Yorkshire dialect that many readers have difficulty with. And Lawrence actually knew this dialect intimately himself. Actually, also in addition to using the dialect in Mellor's case, he wrote a number of poems in this dialect form, and those poems have often been compared to the poems of Robert Burns that use a Scottish dialect in powerful ways.

Well, Lawrence's sexual explicitness, for which he in some ways is most scandalously infamous, has not worn as well as Joyce's in some respects. My students often laugh at Constance Chatterley wrapping Mellor's genitals with flowers to celebrate her sexual awakening. Maybe I should amend that, because I haven't taught the novel to undergraduates in more than 10 years; partly because I found their reaction so unhelpful, and I had to spend so much time trying to say, well, look, you had to think back to what it was like before you could find pornography by clicking the mouse, before every magazine on the newsstand showed nakedness or near nakedness. You have to try to think yourself back to a time when acknowledging the body in these ways was truly shocking, was truly pioneering, and it was difficult for contemporary students to understand that.

So, they found Lawrence's incantatory, passionate language laughable, maybe even naïve. Maybe they were partly right. Maybe Lawrence's most lasting fiction is in his short stories, where the same tendencies in his work are present but more rigorously controlled and confined. As I said, we'll look at one of the greatest of these stories, "The Horse Dealer's Daughter," in the next lecture.

Lawrence once thought to title *Lady Chatterley's Lover* "Tenderness." This sounds strange, I suppose, or might seem strange. But it may not be so strange if we can, for a moment, shed that mass-media-induced cynicism I've been mentioning toward sexual and scatological material. Is that possible in today's world? I'm not sure. But can you hear the tenderness in this passage, a passage that has mobilized or generated a great deal of amusement and mockery from readers and from critics? After lovemaking, the gamekeeper and Lady Chatterley lie naked together. This is from chapter 15 of *Lady Chatterley's Lover*:

> He stroked her tail with his hand, long and subtly taking in the curves and the globe-fullness. "Tha's got such a nice tail on thee," he said, in the throaty caressive dialect. "Tha's got the nicest arse of anybody. It's the nicest, nicest woman's arse as is! An' ivery bit of it is woman, woman sure as nuts. Tha'rt not one o' them button-arsed lasses as should be lads, are ter! Tha's got a real ... sloping bottom on thee, as a man loves in 'is guts. It's a bottom as could hold the world up, it is!"
>
> All the while he spoke he exquisitely stroked the rounded tail, till it seemed as if a slippery sort of fire came from it into his hands. And his finger-tips touched the two secret openings of her body, time after time, with a soft little brush of fire.
>
> "An' if tha shits an' if tha pisses, I'm glad. I don't want a woman as couldna shit nor piss."

Maybe I should interrupt and put in a parenthesis here. There's a famous 18th-century poem by Jonathan Swift making fun of the literary conventions and the falsifications of certain forms of court poetry, and the poem begins, "Celia, Celia, Celia shits." And Lawrence, commenting on the poem at one point with great contempt said, "Of course Celia shits. How could it be otherwise? Who would want it otherwise?" He says this in one of his letters.

> "An' if tha shits an' if tha pisses, I'm glad. I don't want a woman as couldna shit nor piss."

> Connie could not help a sudden snort of astonished laughter
> …

And I think maybe readers also feel like giving a snort of astonished laughter. It's part of Lawrence's genius, I think, often to see through his own comic impulses. He dramatized himself over and over again in his novels, but he dramatized himself in a way that was very unflattering, that could see through his own pomposity and his didacticism, and it's one of the things that makes his novels so powerful, that the D. H. Lawrence characters in his novel are often the characters who are subjected to the most withering criticism and mockery.

> Connie could not help a sudden snort of astonished laughter, but he went on unmoved.
>
> "Tha'rt real, tha art! Tha'art real, even a bit of a bitch. Here tha shits an' here tha pisses: an' I lay my hand on 'em both an' like thee for it. I like thee for it. Tha's got a proper, woman's arse, proud of itself. It's none ashamed of itself."
>
> He laid his hand close and firm over her secret places, in a kind of close greeting.

Well, if we're closed to this totally, if we can't hear the wonder and the tenderness in these maybe comic and partly grating lines, we won't understand one central reason the Modernists really mattered, how they changed literature, and our ideas of ourselves.

Joyce does the sex part better, I think, with less religious fervor than Lawrence, but both are comrades in the great intellectual struggle to open literature to the full range of human experience, and in particular to accept and explore the great recognition that Yeats expresses unforgettably in his poem "Crazy Jane Talks to the Bishop": "Love has pitched his mansion in the place of excrement."

Lecture Twelve
"Horse Dealer's Daughter"—A Shimmer Within

Scope:

Romanticism and Modernism had once been seen as discontinuous movements, having little in common. But in recent decades, literary scholarship has recognized a continuity between the two. Romanticism dwells upon issues of consciousness and selfhood. In this sense, D. H. Lawrence is a direct descendant of many Romantic poets. We look at one scene from Lawrence's *Women in Love* that dramatizes his sense of the fluidity and volatility of human nature. Then we turn to his short story "The Horse Dealer's Daughter," which shows Lawrence at his signature best, aiming to evoke energies and feelings that are beyond or beneath language and rational thought.

Outline

I. We begin this second lecture on D. H. Lawrence by talking briefly about the relationship between Modernism and Romanticism, the great movement in art and literature in the first part of the 19th century.

   A. Scholars have come to view Romanticism and Modernism as sharing certain characteristics.

   B. English Romantic poetry in particular prefigures Modern fiction's preoccupation with ordinary experience and with nature.

   C. In addition, Romanticism is preoccupied with problems of consciousness and selfhood, and it is particularly in this realm that Lawrence's Romantic heritage is clear.

   D. Lawrence focuses on elements that lie beneath the surface and are, in some cases, inexpressible or challenge the possibilities of language.

   E. He is obsessed by themes of rebirth.

   1. Lawrence's imagery suggests a rebirth that requires ending confinement.

2. Much of his imagery is drawn from nature and aims to break through the accumulated, mechanical operations of our social lives to energies and instincts long suppressed.

II. One scene from Lawrence's *Women in Love* dramatizes this theme of breaking free and also discloses his keen awareness of the unstable volatility of human nature.

   A. The passage comes from the chapter titled "Excurse," in which Ursula and Birkin vehemently argue.

   B. Ursula heaps vituperation on Birkin, who recognizes the truth of what she says. After enduring Ursula's tremendous verbal abuse, Birkin "…stood watching in silence. A wonderful tenderness burned in him, at the sight of her quivering, so sensitive fingers: and at the same time he was full of rage and callousness." The conflicting feelings show Lawrence's understanding of the contradictory and flammable nature of human emotions.

   C. Ursula continues to criticize Birkin for being a "truth-lover" and "purity-monger," terms that also describe Lawrence himself, who was famous for his self-righteousness. This tells us something about Lawrence's power of self-criticism.

   D. The battle continues, intensifying before Ursula runs off and Birkin remains, with "…a darkness over his mind."

   E. Ursula returns with a flower for Birkin, who now feels "…sweetly at ease, the life flowed through him as from some new fountain, he was as if born out of the cramp of a womb." Then they declare their happiness in a sudden reversal of explosive emotions.

   F. The emotions are so intense that they liberate both characters into a new freedom and tenderness. Here, Lawrence expresses the idea that the self is always in process; extreme emotions can be dangerous, but once one breaks through them, they become enabling.

   G. Although he never claims full omniscience, Lawrence adapts a strategy of the 19th-century tradition of omniscient narrators. In many cases, he even reports what his characters feel before they themselves are fully aware of their emotions.

1. This approach allows him to treat submerged energies in a way that no other literary strategy permits.
2. In preserving a vestige of the powers of omniscient narrator, Lawrence pursues the Modernist project of uncovering instinctual energies that he believes modern civilization has denied.

III. "The Horse Dealer's Daughter" (1922) resides at the center of Lawrence's achievement. The story captures the "shimmer within" that lies beyond or beneath language.

   A. "The Horse Dealer's Daughter" is Mabel Pervin; bereft of purpose in life, facing eviction and without prospects, she is apparently destined to become a domestic servant.
   B. Mabel tries to drown herself but is rescued by Dr. Fergusson. The two realize their love for each other but only after resisting it.
   C. The characters seem unmotivated in many ways. Why are they drawn to each other, and why do they resist falling in love?
   D. Mabel has been the housekeeper for the ten years following her mother's death. Now that her father has died, the farm will be sold and her brothers will leave to seek other work.
   E. Mabel has no prospects. Her identity has been stripped away, and her home is about to be taken away. She is a sullen and deeply inarticulate character, sometimes described in animal terms.
   F. When she walks into the pond to drown herself, the doctor pulls her out, even though he cannot swim. Lawrence's rich description of Fergusson sinking into the stinking clay of the pond, then emerging with Mabel, hints at a baptism and transformation.
   G. Fergusson brings her to his house, where he undresses and cleans her. Once Mabel regains consciousness, she asks Fergusson if he loves her. Coming out of the water has awakened energies in her that she herself did not know existed.
   H. She grabs his legs and begins passionately kissing his knees. Fergusson "…was amazed, bewildered and afraid. He had

never thought of loving her. He had never wanted to love her." He is not attracted to her, yet "…he had not the power to break away."

    I. By the story's conclusion, Fergusson begins to acknowledge the feelings that emerge against his conscious understanding and perhaps even against his will. Eventually, he recognizes his love for Mabel.

IV. In "The Horse Dealer's Daughter," Lawrence suggests that to embark on love is dangerous, even frightening.

    A. The story ends on a complex note of both triumph and fear, in which the characters are newly awakened to previously unacknowledged feelings and emotions.

    B. In his last published work, *Apocalypse and the Writings on Revelation* (published posthumously in 1931), he expresses his sense of our human vitality on more time:

> For man, the vast marvel is to be alive. For man, as for flower and beast and bird, the supreme triumph is to be most vividly, most perfectly alive. Whatever the unborn and the dead may know, they cannot know the beauty, the marvel of being alive in the flesh. The dead may look after the afterwards. But the magnificent here and now of life in the flesh is ours, and ours alone, and ours only for a time. We ought to dance with rapture that we should be alive and in the flesh, and part of the living, incarnate cosmos. I am part of the sun as my eye is part of me. That I am part of the earth my feet know perfectly, and my blood is part of the sea.

Essential Reading:

Lawrence, "The Horse Dealer's Daughter," in *The Complete Short Stories of D.H. Lawrence* or *Short Story Masterpieces*, Warren and Erskine, eds.

Supplementary Reading:

Lawrence, *Women in Love*.

Questions to Consider:

1. What verbal strategies does Lawrence use to dramatize unconscious impulses and energies, and why are such energies important to him?
2. What is the significance of Mabel's attempt to drown herself and of Fergusson's near-death when he rescues her from the muddy dark of the pond?

Lecture Twelve—Transcript
"Horse Dealer's Daughter"—A Shimmer Within

I want to begin this second lecture on D. H. Lawrence by talking briefly about the relationship between Modernism and Romanticism, the great movement in art and literature in the first part of the 19th century, a kind of counterpart in a way to the Modernism that emerges in the last years of the 19th and the first years of the 20th century.

It used to be thought, and it was kind of commonplace of scholarship and criticism when I was an undergraduate, that Romanticism and Modernism were enemies, had very little in common with each other. But in the years that followed, and certainly over the last 20 years or more, literary scholarship has come to more and more recognize that there's a kind of continuity between that first revolution of the word, Romanticism, and Modernism, and that while the continuity is an imperfect one—it isn't as if there's a perfect line between them—there are many areas of connection and overlap.

Romantic poetry, especially English Romantic poetry, prefigures our Modern fiction writers' preoccupation with ordinary experience. We might think, for example, of the great English poet Wordsworth's interest in figures like leech gatherers and solitary reapers. One of Wordsworth's great poems begins, "Behold her singing in the field, yon solitary lass" (I'm paraphrasing) and the whole poem is about the sort of solitary work of this reaper who's working in the fields.

And Romanticism is also particularly preoccupied, as Harold Bloom and other Romantic scholars have emphasized, with problems of consciousness and selfhood, with the mysteries within, with our passionate lives. And in that sense especially D. H. Lawrence is a direct descendant of many of the Romantic poets.

Conrad has a Romantic aspect as well and some of our other writers do in addition, but Lawrence may be the writer with the strongest link to this earlier literary revolution. We might call Lawrence a Romantic vitalist and link him especially to the apocalyptic Romanticism of poets like William Blake. I'll try to indicate in what way this is significant a bit later on, but I can anticipate my argument by simply reminding you of the extent to which Lawrence is constantly preoccupied by some energies or elements that are beneath the surface, that are in some cases inexpressible, or challenge the possibilities of language—that aren't even visible on the surface, but have a kind of authority for him.

We might think of the following as D. H. Lawrence's mission statement. This is a passage from the novel I mentioned earlier, *Sons and Lovers*, and in this passage Paul Morel, the would-be painter figure, the would-be artist in the novel, is having a conversation with his girlfriend. And she is very sensitive and sympathetic to his work, and in fact in many ways is very nourishing for him. Part of the tragedy of the novel, in fact, is that she seems such a wonderfully appropriate partner for him, but he's incapable of a long-term commitment to her. In any case, this is a scene between them, and she's looking at his paintings or his sketches, and she says:

> "Why do I like this [sketch] so?" …
>
> "Why DO you?" he asked.
>
> "I don't know. It seems so true," [she says.]

And then Paul answers:

> "It's because—it's because there is scarcely any shadow in it; it's more shimmery, as if I'd painted the shimmering protoplasm in the leaves and everywhere, and not the stiffness of the shape. That seems dead to me. Only this shimmeriness is the real living. The shape is dead crust. The shimmer is inside really."

"The shimmer is inside really." Lawrence's imagery, whether he's writing poetry or fiction, is full of moments in which individuals or sometimes creatures break free of carapaces, break free of integuments, shatter chrysalises, come to a kind of birth through difficulty that requires the breaking of shells, the breaking of surfaces, the breaking of enclosures and confinements. And of course he draws a lot of this imagery from the natural world, and he writes poems about things like caterpillars turning into butterflies; but that natural process offers him a kind of fundamental way in to his sense of human experience, and to his sense of those qualities in human life that have been caulked over or balked or suppressed or denied by the accumulations of civilized responsibility, by the accumulations that follow from the demands that are made upon us by social occasions, by our jobs, by our commitment to certain forms of propriety and manners. So Lawrence is obsessed by moments that involve a kind of birthing or a coming to life.

And there are a number of places in his work where this becomes particularly clear, in which these impulses are especially powerfully crystallized or distilled. And I want to call your attention to one of them now because it will not only further the idea, clarify this idea, and help us to understand one of the central points, maybe *the* central idea of "The Horse Dealer's Daughter," to which I want to devote most of our attention in this lecture, but also because it can illuminate some of the ways in which Lawrence is to be distinguished from writers like Ford Madox Ford or Conrad.

The scene I want to talk about is a scene from *Women in Love*, and it's in a chapter titled "Excurse," and it's a wonderful memorable chapter. I sometimes have students read it even when I do not assign the novel as a whole to them, in part because it dramatizes something that seems to me so true about human relationships, but so rarely dramatized by writers—partly because it's so difficult to dramatize.

It's a scene between Ursula and Birkin (two of the four central characters in the novel I spoke about last time), and it's a moment in their fraught and difficult relationship, which goes through many phases as the novel continues, in which they are at loggerheads. They have an immense, an intense, a shocking fight, a quarrel so intense that you half expect one or the other of them to strike their partner.

And in the scene you can get some sense of the intensity and the power of Lawrence's ability to dramatize dialogue and to dramatize things like quarrels, from this sequence. The quarrel goes on for a number of pages, and Ursula heaps vituperation upon Birkin, and one of the most moving and interesting parts of the scene is the extent to which Birkin recognizes in some degree the truthfulness of what she's saying; and after suffering or enduring a tremendous amount of abuse from Ursula, Lawrence writes this sentence of Birkin:

> He stood watching in silence. A wonderful tenderness burned in him, at the sight of her quivering, so sensitive fingers: and at the same time he was full of rage and callousness.

Well the contradictory feelings that are dramatized there, that he feels a wonderful tenderness burning in him—and the idea that tenderness might burn is a particularly Lawrentian idea—but that he can feel tenderness at the same moment that he feels full of rage and callousness is a characteristic insight of Lawrence's, having to do with how deeply contradictory and volatile human emotions are.

And then he says to her, "This is a degrading exhibition." And she says, "Yes, degrading indeed, but more to me than to you."

And a few lines later Ursula says: "YOU!" she cried. "You! You truth-lover! You purity-monger!" One of the wonderful things about these passages is they magnificently describe not only the hectoring, lecture-oriented Birkin that we meet in the novel, but they also, of course, describe D. H. Lawrence himself, who was famous for his righteous and self-righteous energy. He was a kind of prophet, saw himself in these terms and was full of a kind of moralistic condemnation and contempt for what he felt were the evasions and lies of conventional experience.

So when she's attacking Birkin here she's also, of course, articulating a critique of D. H. Lawrence himself; and that it is from the pen of D. H. Lawrence tells us something about his power of self-criticism. She cries:

> "You truth-lover! You purity-monger! It STINKS, your truth and your purity. It stinks of the offal you feed on, you scavenger dog, you eater of corpses. You are foul, FOUL, and you must know it. Your purity, your candour, your goodness—yes, thank you, we've had some. What you are is a foul, deathly thing, obscene, that's what you are, obscene and perverse."

And Birkin stands there and takes it. And the battle goes on, and the energy between them becomes, if anything, even more intense, in some ways even more terrible, and finally the scene concludes in this way:

> There was a darkness over his mind. The terrible knot of consciousness that had persisted there like an obsession was broken. ... [Listen for the imagery of the breaking of integuments, the breaking into a new kind of freedom or understanding.] He wanted her to come back. [She's run off; she's gone away.] He breathed lightly and regularly like an

infant, that breathes innocently, beyond the touch of responsibility ... He was as if asleep, at peace, slumbering and utterly relaxed.

She came up and stood before him ... "See what a flower I found you," she said.

... His mind was sweetly at ease, the life flowed through him as from some new fountain, he was as if born out of the cramp of a womb.

"Are you happy?" she asked ...

"Yes," he said.

"So am I," she cried in sudden ecstasy.

"'So am I,' she cried in sudden ecstasy." Well, there are many things about this astonishing passage that continue to surprise and amaze me whenever I read it. One of them, of course, is the intensity of the feeling that is expressed here, but another is the surprising reversal that occurs within the scene, as if the explosion of emotion is so intense and so complete that it liberates both of them into a new kind of freedom and into a new kind of tenderness.

One of the things that Lawrence is dramatizing here is the volatility of emotions, the volatility of the self itself; the idea that the self is in some sense unstable, always in process, open to extremes of response, extremes of emotion that can be immensely dangerous, but also, once one breaks through them, enabling emotions as well, is part of what Lawrence is up to here.

It's worth noticing that there is a great resemblance between this passage and the way in which Virginia Woolf, in the novel we'll talk about later on in these lectures, *To the Lighthouse*, deals with the self and with the inner sense. Because Woolf's novel also is conscious of how unstable our inner life is, how volatile it is, how extreme the emotions are that roil around submerged beneath the level of consciousness or conscious awareness.

And although Lawrence and Virginia Woolf are often thought of as very different writers, in certain fundamental respects they share a sense of the immense energy, even the murderous and extreme energy, that is bottled up within us, and that finds expression sometimes in quarrels, sometimes in dreams, sometimes simply in resentful and angry thoughts, or in tender thoughts.

So, this scene from *Women in Love* has a great authority and power as a kind of emblem for the way Lawrence understands the adventure of human relationships. It's also important as well, because it tells us something about Lawrence's practice as a writer. If you think about the passages that I have read and paraphrased you'll realize, I think, that Lawrence behaves in some respects not like Conrad or Ford, who give us first-person narrators whose perspective on the world is limited to their own angle of vision, but that Lawrence adapts a strategy that belongs more to the 19th century tradition of the omniscient narrator.

Lawrence never claims the full omniscience, the grand omniscience of a Tolstoy or a George Eliot, but he does claim the right as a fiction writer to delve into the minds and beneath the minds, into the emotions, into the instinctual depths, of his characters and to report what they're feeling. In many cases to report what they're feeling before the characters themselves are even fully aware of it.

In his adaptation of this omniscient strategy, Lawrence is able to talk about those submerged energies in a way that no other literary strategy would permit. And it is, I think, a mark of his conservatism, in a certain sense, that he preserves at least this vestige of the powers of the omniscient narrator in order to pursue what is in the end a very Modernist project, the project of uncovering these instinctual energies that he believes have been submerged or covered up or denied by modern civilization.

Well, "The Horse Dealer's Daughter" dramatizes these kinds of things with a particular austerity and intelligence. Many would say that Lawrence's most lasting art is in the short story because the compression that it creates brings out his genius in a distilled way, in a way that minimizes excess. In any case, "The Horse Dealer's Daughter" certainly resides at the very center of Lawrence's achievement. It has a working-class theme. It shows the kind of accuracy of setting, place, and physical description of ordinary life and of English working-class environments that are characteristic of the very best Lawrence stories. And the central energies of the story point inward in an effort to capture those inner energies I've been talking about: that shimmer, what is beyond or beneath words.

A basic plot summary: Outwardly the story is a kind of romance magazine tale. A boy meets girl; the boy gets the girl. Bereft of purpose in life, facing eviction for bankruptcy from her family home and business, without prospects, and apparently fated for the humiliation of becoming a maid or some kind of domestic—where before she had run the household in place of her dead mother—the central character of the story, Mabel Pervin, visits her mother's grave, then walks into a pond attempting to drown herself.

She's rescued by a doctor named Ferguson, who revives her—takes her clothes off while she's unconscious, removes her clothing, dries her off, brings her down by the fire, and then she revives. And in a strangely echoing, hammering climax, the two characters realize a love for each other that they hadn't even recognized had existed before. They resist and fear this recognition, and then embrace it as the story ends.

The characters seem unmotivated in many ways, and when I teach this story to my undergraduates, the students are often utterly puzzled by why the characters are drawn to each other, and especially by the question of why the characters seem to resist and to fear falling in love. Well, even the characters in the story don't understand exactly why they're doing what they're doing. What D. H. Lawrence is doing in the story is aiming for what could be called the shimmer inside. He wants to describe the working of desires, of energies beneath consciousness, maybe that are too frightening for consciousness.

But he does plant some realistic details in the story. The story opens with the doctor visiting Mabel's household. She and her brothers are sitting around the table on one of the last days in which they can remain in their household. Their father has died and left them in debt. The horses that had been their living have all been sold off, and their homestead is being sold off.

The brothers are going off to work in various places, and Mabel is described as particularly sullen and unhappy, almost in animal terms. She's described as a kind of sullen animal, a sullen creature who won't even acknowledge what her brothers ask her. Her brothers say, What are you going to do, Mabel? Are you going to become a domestic? Are you going to go and live with your sister? But she won't answer.

The story tells us that Mabel for ten years had been taking care of the family, since her mother's death; that had given her a kind of identity, a kind of pride in function that has now been completely removed. It's as if her identity is gone. But she's a sullen and deeply inarticulate character.

When the doctor arrives to sort of visit with the brothers—he's clearly the friend of one of the brothers—he addresses the brothers by their first names, but he addresses Mabel by Miss Pervin, and the implication is that they are not close. But Lawrence plants a minor detail there, saying that her eyes had always unsettled him; and the implications of that come clear later in the story.

The doctor goes about his daily tasks of caring for people, and at a certain point coming out of his surgery, he looks toward the graveyard, to the village graveyard, and he sees Mabel Pervin presumably paying a visit to her mother's grave. And then he sees her walk off into a pond as if to drown herself and he rushes after her, and there's a remarkable scene where the doctor rescues her. She's literally attempting suicide. She thinks she has no purpose left in life.

And Lawrence describes his going to the pond. He doesn't know how to swim, and the pond is not so deep that he needs to swim out to her, but if he did go under the water he would be in some danger. And Lawrence describes this moment. It's really a kind of moment of rebirth, a kind of resurrection moment in which the immersion in the pond comes to stand for a kind of baptism or a kind of rebirth, a visitation or a touching of the very bottom of life; a hint that both Mabel and the doctor who tries to rescue her come close to death and then emerge from it resurrected, changed, transformed.

Here's a passage from the story: "He slowly ventured into the pond. The bottom was deep, soft clay, he sank in, and the water clasped dead cold round his legs ... He could smell the cold, rotten clay that fouled up into the water." Listen to Lawrence's habit of repetition, which is an aspect of the incantatory quality of his prose.

> He could smell the cold, rotten clay that fouled up into the water. It was objectionable in his lungs. ... The lower part of his body was all sunk in the hideous cold element. ... He lost his balance and went under, horribly, suffocating in the foul earthy water. ... At last, after what seemed an eternity, he

got his footing, rose again into the air …. He gasped, and knew he was in the world. [A kind of rebirth, right?] Then he looked at the water. She had risen near him. [She had *risen* near him.] He grasped her clothing, and drawing her nearer, turned to take his way to land again. … He lifted her and staggered on to the bank, out of the horror of wet, grey clay.

Think how often words like "clay" or "cold" or "dead" are repeated in the passage. He then takes her home, as I've mentioned, and when she awakens, she looks up at him in a kind of amazement and wonder and is astonished. She says, "Do you love me, then?" "He only stood and stared at her, fascinated. His soul seemed to melt," Lawrence says. Lawrence is one of the very few Modern writers who would write a sentence like that. "His soul seemed to melt." He doesn't mean soul in a religious sense, although he is in a certain way a kind of religious writer, a vitalist. His religion is the religion of Romantic vitalism.

> She shuffled forward on her knees, and put her arms round him, round his legs, as he stood there, pressing her breasts against his knees and thighs, clutching him with strange, convulsive certainty, pressing his thighs against her, drawing him to her face, her throat, as she looked up at him with flaring, humble eyes of transfiguration, triumphant in first possession.

It's as if her coming out of the water and her awakening here has awakened energies in her that she herself didn't even know existed.

> And she was passionately kissing his knees, through the wet clothing, passionately and indiscriminately kissing his knees, his legs, as if unaware of everything.

And his reaction is almost comic. At first he's horrified. He doesn't acknowledge that he has any feelings for her. It says: "He was amazed, bewildered and afraid. He had never thought of loving her. He had never wanted to love her," Lawrence writes:

> When he [had] rescued her and restored her, he was a doctor, and she was a patient. He had … no single personal thought of her. Nay, this introduction of the personal element was very distasteful to him, a violation of his professional honour. It was horrible to have her there embracing his

knees. It was horrible. He revolted from it, violently. And yet—and yet—he had not the power to break away.

And then in the finale of the story we see him beginning to acknowledge the feelings that lie deep within him, that begin to emerge against his conscious understanding, and maybe even against his conscious will. And of course he comes to feel that he loves her as well.

And the story rises to a kind of violent climax in which both characters feel at least as much fear and uncertainty about this new affection, this new love that they feel for each other as they do excitement. As if Lawrence is suggesting that to embark on the risk of love and to open yourself to the energies of love, as he defines love—which is a deep and instinctual thing, not a matter of consciousness at all—that to embark on such an adventure is fraught with danger as well as excitement.

And so the story ends on a complex note, of triumph but also fear, in which both characters are newly awakened in a certain sense to new feelings and to new emotions, which they had not even acknowledged existed before.

This kind of thing is characteristic of Lawrence at his best, in which he tries to dramatize characters and the experience of characters who themselves don't understand what is motivating them. This is an immensely difficult thing, of course, for a writer to do, and it's often an immensely difficult thing for a reader to follow.

When my students read this story, as I mentioned, they're often deeply puzzled. They first don't understand why the characters would be drawn to each other, and once they see that Lawrence's incantatory prose aims to press toward a revelation of the inexpressible, or a revelation of something that is beyond words, they can begin to recognize that. But then once the characters become aware of and begin to acknowledge their feelings, students are troubled by the idea that the characters would be so uneasy about them; but that's partly because they're young, I think, and it's partly because they have not yet fully understood that deep feelings, that deep and profound emotions are dangerous, are demanding, are fraught with difficulty, and are problematic, test us in terrifying and powerful ways.

One way to describe what I've been talking about and to characterize Lawrence's vision of life is to say that Lawrence thinks of life as a deed, as a constant act in which we must choose to face what's scary and dangerous; that life is an ongoing action. We don't need elaborate theories, I think, to feel the truth of Lawrence's sense of life as a lovely, painful drama of enmity and struggle, the kind of thing that's dramatized so powerfully in "The Horse Dealer's Daughter," in which two characters who didn't even know they were destined for each other discover their destiny together, discover their love together, and discover that that love is dangerous and exciting—exhilarating, but also dangerous. We don't need elaborate theories to feel the truth of this.

His best work, I think, is about aliveness, how hard it is to keep hold of. He's the van Gogh of literary Modernism; those garish, desperate colors and slashes of thick paint are the visual emblem of Lawrence's incantatory prose. He wants to get at the vitality that's surging against what's confining or repressing it.

Lawrence said this again and again in his fiction and poetry. In his last published work, published posthumously in 1931, he said it one last time. And I want to end with this quotation from a text that was published under the title *Apocalypse*:

> For man, [Lawrence wrote,] the vast marvel is to be alive. For man, as for flower and beast and bird, the supreme triumph is to be most vividly, most perfectly alive. Whatever the unborn and the dead may know, they cannot know the beauty, the marvel of being alive in the flesh. The dead may look after the afterwards. But the magnificent here and now of life in the flesh is ours, and ours alone, and ours only for a time. We ought to dance with rapture that we should be alive and in the flesh, and part of the living, incarnate cosmos. I am part of the sun as my eye is part of me. That I am part of the earth my feet know perfectly, and my blood is part of the sea.

Timeline

1880 ....Fyodor Dostoyevsky's *The Brothers Karamazov*; Henry James's *Washington Square*; death of Gustave Flaubert, George Eliot

1881 ....Henry James's *The Portrait of a Lady*; U.S. President Garfield assassinated; birth of Pablo Picasso

1882 ....Beginning of psychoanalysis as Joseph Breuer uses hypnosis to treat hysteria; Maxim machine gun patented; birth of Franklin Roosevelt, Virginia Woolf, James Joyce, Igor Stravinsky

1883 ....Friedrich Nietzsche's *Also sprach Zarathustra*; first 10-story skyscraper in Chicago; opening of the Brooklyn Bridge; death of Karl Marx; birth of Benito Mussolini, Franz Kafka, John Maynard Keynes

1884 ....Mark Twain's *Huckleberry Finn*; *Oxford English Dictionary* begins publication

1885 ....Death of General Gordon defending British possession of Khartoum from the Mahdi in Sudan; Emile Zola's *Germinal*; Gilbert and Sullivan's *The Mikado*; first photographic paper manufactured by George Eastman; birth of D. H. Lawrence, Ezra Pound

1886 ....Robert Louis Stevenson's *Dr. Jekyll and Mr. Hyde*; Arthur Rimbaud's *Illuminations*; Richard von Krafft-Ebing's *Psychopathia Sexualis*; Nietzsche's *Beyond Good and Evil*; Statue of Liberty dedicated; Georges

	Seurat finishes *A Sunday on La Grande Jatte*
1887	Arthur Conan Doyle's first Sherlock Holmes story, "A Study in Scarlet"; Nietzsche's *On the Genealogy of Morals*; invention of celluloid film; birth of Marc Chagall, Le Corbusier
1888	Rudyard Kipling's *Plain Tales from the Hills*, which includes *The Man Who Would Be King*; Jack the Ripper murders six women in London; birth of T. S. Eliot, Eugene O'Neill, Irving Berlin
1889	Eiffel Tower designed for the Paris World Exhibition; André Gide begins publication of *Journal*; birth of Adolf Hitler, Charles Chaplin, Jean Cocteau, Martin Heidegger
1890	Thomas Edison shows first motion pictures in New York City; Henrik Ibsen's *Hedda Gable*; Oscar Wilde's *The Picture of Dorian Gray*; J. G. Frazer's *The Golden Bough*; death of Vincent van Gogh
1891	Invention of the clothing zipper; Gustav Mahler's *Symphony No. 1*; death of Herman Melville, Arthur Rimbaud
1892	Claude Monet begins a series of paintings of Rouen Cathedral; first internal combustion engine patented; Peter Tchaikovsky's *The Nutcracker*; death of Walt Whitman, Alfred Lord Tennyson
1893	Beginnings of Art Nouveau in Europe; Henry Ford builds his first car; World Exhibition in Chicago

1894	Arrest and deportation of Captain Alfred Dreyfus in France; first gramophone disk in Germany; Rudyard Kipling's *The Jungle Book*; Claude Debussy's *L'Après-midi d'un faune*; birth of Aldous Huxley
1895	Lumière brothers invent Cinématographe, a motion-picture camera and projector; Cuba begins its struggle for independence from Spain; Oscar Wilde sues the marquis of Queensberry for libel; H. G. Wells's *The Time Machine*; birth of Lewis Mumford
1896	Establishment of Nobel Prize; first modern Olympics in Athens; Giacomo Puccini's *La Bohème*; Anton Chekhov's *The Sea Gull*
1897	First American comic strip, *The Katzenjammer Kids*, begins publication; discovery of the electron by J. J. Thomson; Diamond Jubilee of Queen Victoria
1898	United States declares war on Spain over Cuba; opening of the Paris Metro; birth of George Gershwin, Bertolt Brecht, Ernest Hemingway
1899	First publication of Joseph Conrad's *Heart of Darkness* in *Blackwood's Magazine*; first sound recording using magnetic medium; Scott Joplin composes "Maple Leaf Rag"; Oscar Wilde's *The Importance of Being Earnest*; birth of Vladimir Nabokov, Federico Garcia Lorca
1900	Boxer Rebellion in China; Conrad's *Lord Jim*; Sigmund Freud's *The*

	Interpretation of Dreams; Max Planck publishes papers on quantum theory; death of Friedrich Nietzsche
1901	Founding of U.S. Steel Corporation by J. P. Morgan; U.S. President McKinley assassinated; Nobel Peace Prize established; Thomas Mann's *Buddenbrooks*; death of Giuseppe Verdi; birth of Walt Disney
1902	First recording of Enrico Caruso; William James's *Varieties of Religious Experience*; Beatrix Potter publishes *Peter Rabbit*; eruption of Mount Pelée in Martinique
1903	First airplane flight by the Wright Brothers; Henry James's *The Ambassadors*; Edwin S. Porter directs *The Great Train Robbery*, often identified as the first significant narrative film; first Tour de France
1904	Beginning of Russo-Japanese War; work on the Panama Canal begins; Conrad's *Nostromo*; Max Weber's *The Protestant Ethic and the Spirit of Capitalism*; Puccini's *Madame Butterfly*; James Barrie's *Peter Pan*
1905	Mutiny on the battleship *Potemkin* during Russo-Japanese War; Einstein's special theory of relativity; F. T. Marinetti's *Futurist Manifesto*; rise of Fauvism in France; first regular cinema established in Pittsburgh
1906	Everyman's Library begins publication in London; Upton

©2007 The Teaching Company.

201

Sinclair's *The Jungle*; death of Paul Cézanne; birth of Samuel Beckett, Greta Garbo

1907 .... Lord Baden-Powell founds the Boy Scouts; Picasso's *Les Demoiselles d'Avignon* inaugurates Cubism; Henri Bergson's *Creative Evolution*; Rudyard Kipling wins Nobel Prize; birth of W. H. Auden

1908 .... Establishment of the Ashcan School of painters in New York City; Béla Bartók's *String Quartet No. 1*; E. M. Forster's *A Room with a View*; Gertrude Stein's *Three Lives*; Ezra Pound moves from the United States to London

1909 .... First abstract paintings of Wassily Kandinsky; first movie newsreels; first commercial production of plastic

1910 .... Postimpressionist Exhibition organized in London by Roger Fry; death of Leo Tolstoy, Mark Twain, Mary Baker Eddy

1911 .... Conrad's *Under Western Eyes*; Arnold Schoenberg's *Manual of Harmony*; theft of *Mona Lisa* from the Louvre

1912 .... Conrad's *The Secret Sharer*; Arizona and New Mexico statehood completes continental United States; Industrial Workers of the World leads textile workers' strike in Lawrence, Massachusetts; sinking of the *Titanic*; founding of F.W. Woolworth Company; birth of Eugène Ionesco

1913 .... Armory Show in New York City introduces European Modernist art to the United States; suffrage demonstrations in London; opening of Grand Central Station in New York City; first Charlie Chaplin movies; D. H. Lawrence's *Sons and Lovers*; Thomas Mann's *Death in Venice*; Marcel Proust's first of seven volumes of *À la recherche du temps perdu*; Stravinsky's *Le Sacre du Printemps*; birth of Benjamin Britten, Albert Camus

1914 .... Beginning of World War I; James Joyce's *Dubliners*; Edgar Rice Burroughs's *Tarzan of the Apes*; opening of the Panama Canal; birth of Tennessee Williams

1915 .... Albert Einstein's general theory of relativity; D. W. Griffith directs *Birth of a Nation*; First Dada paintings of Marcel Duchamp; D. H. Lawrence's *The Rainbow*; Edgar Lee Masters's *Spoon River Anthology*; Franz Kafka's *The Metamorphosis*; Ford Madox Ford's *The Good Soldier*; Ezra Pound begins work on the *Cantos* (completed in 1962); birth of Saul Bellow, Arthur Miller

1916 .... Joyce's *Portrait of the Artist as a Young Man*; John Dewey's *Democracy and Education*

1917 .... October Revolution in Petrograd; Balfour Declaration on Palestine; Conrad's *The Shadow-Line*; first jazz recordings in the United States

1918 .... Conclusion of World War I; first exhibition of Joan Miró; Henry Adams's *The Education of Henry Adams*

1919 .... Founding of Bauhaus in Germany by Walter Gropius; establishment of the Radio Corporation of America; Sherwood Anderson's *Winesburg, Ohio*; Benito Mussolini founds the Fascist Party in Italy

1920 .... Lawrence's *Women in Love*; League of Nations established; 18th Constitutional Amendment prohibits alcohol in the United States; 19th Amendment grants American women voting rights; Kafka's "A Country Doctor"; Sinclair Lewis's *Main Street*; Edith Wharton's *The Age of Innocence*; first performance of Gustav Holtz's *The Planets*; Robert Weine directs *The Cabinet of Dr. Caligari*; Herman Rorschach devises his "inkblot" test; death of John Reed, Max Weber

1921 .... Luigi Pirandello's *Six Characters in Search of an Author*; Ludwig Wittgenstein's *Tractatus-Logico-Philosophicus*

1922 .... Joyce's *Ulysses*; Lawrence's "The Horse Dealer's Daughter"; T. S. Eliot's *The Waste Land*; Sinclair Lewis's *Babbitt*; tomb of Pharaoh Tutankhamen discovered in Egypt; F. W. Murnau directs *Nosferatu*; *Reader's Digest* established in the United States; death of Marcel Proust

1923 .... Great Tokyo earthquake kills 120,000; William Butler Yeats wins Nobel Prize; Isaac Babel's *Odessa Stories*; E. E. Cummings's *The Enormous Room*; Martin Buber's *I and Thou*; George Gershwin's *Rhapsody in Blue*

1924 .... E. M. Forster's *A Passage to India*; Thomas Mann's *The Magic Mountain*; Ezra Pound moves to Italy, supports Mussolini and the Fascist Party; Louis de Broglie's study of the wave theory of matter; J. Edgar Hoover appointed director of the FBI; death of Joseph Conrad, Franz Kafka, V. I. Lenin, and Giacomo Puccini

1925 .... Hitler's first volume of *Mein Kampf*; Virginia Woolf's *Mrs. Dalloway*; Babel's "The Story of My Dovecote"; Theodore Dreiser's *An American Tragedy*; F. Scott Fitzgerald's *The Great Gatsby*; Hemingway's *In Our Time*; Kafka's *The Trial* published posthumously; *New Yorker* magazine established; Sergei Eisenstein directs *Potemkin*; Chaplin directs *The Gold Rush*; anti-evolution Scopes trial held in Tennessee; Dmitri Shostakovich's *Symphony No. 1*; death of Erik Satie

1926 .... Book-of-the-Month Club established; Babel's *Red Cavalry*; Hemingway's *The Sun Also Rises*; R. H. Tawney's *Religion and the Rise of Capitalism*; Fritz Lang directs *Metropolis*; first jazz recordings of Duke Ellington, Jelly Roll Morton released; Robert

Goddard's first liquid-fuel rocket experiments; death of Rainer Maria Rilke, Rudolph Valentino, Claude Monet

1927 .... Leon Trotsky expelled from the Communist Party; Hermann Hesse's *Steppenwolf*; Sinclair Lewis's *Elmer Gantry*; B. Traven's *The Treasure of the Sierra Madre*; Woolf's *To the Lighthouse*; first "talkie" film, *The Jazz Singer*; Leon Theremin invents the first electronic musical instrument; Charles Lindbergh pilots the *Spirit of St. Louis* solo across the Atlantic; Sacco and Vanzetti executed

1928 .... First Five-Year Plan in the U.S.S.R.; Lawrence's *Lady Chatterley's Lover*; Woolf's *Orlando*; Disney's *Steamboat Willie*, the first Mickey Mouse film; Maurice Ravel's *Bolero*; Kurt Weill's *The Threepenny Opera*; Alfred Hitchcock's first sound film, *Blackmail*

1929 .... U.S. Stock Exchange collapse inaugurates worldwide Great Depression; Thomas Mann wins Nobel Prize; Woolf's *A Room of One's Own*; Alfred Döblin's *Berlin Alexanderplatz*; William Faulkner's *The Sound and the Fury*; Thomas Wolfe's *Look Homeward, Angel*; Heidegger's *What Is Philosophy?*; Museum of Modern Art opens in New York City; construction of the Empire State Building begins in New York City; St. Valentine's Day Massacre in Chicago

1930	Faulkner's *As I Lay Dying*; Robert Musil's first volume of *The Man Without Qualities*; Dashiell Hammett's *The Maltese Falcon*; Freud's *Civilization and Its Discontents*; the "new planet" Pluto is discovered; Joseph von Sternberg directs *The Blue Angel*; Grant Wood paints *American Gothic*; death of D. H. Lawrence, Arthur Conan Doyle
1931	Salvador Dali paints *The Persistence of Memory*; completion of the Empire State Building and George Washington Bridge in New York City; death of Thomas Alva Edison
1932	Franklin D. Roosevelt elected president; Britain declares Indian National Congress party illegal, arrests Mahatma Gandhi; Huxley's *Brave New World*; Louis-Ferdinand Céline's *Journey to the End of the Night*; Faulkner's *Light in August*; Fritz Lang directs *M*; kidnapping of Lindbergh baby; death of Hart Crane
1933	Hitler becomes chancellor of Germany; Reichstag burns; first concentration camps established; Tennessee Valley Authority established in the United States; Garcia Lorca's *Blood Wedding*; Mann's first volume of *Joseph and His Brothers*; censorship of Joyce's *Ulysses* ends in the United States (ruling came in 1933, decision affirmed 1934); Carl Jung's *Modern Man in Search of a Soul*; A. N. Whitehead's *Adventures of Ideas*

1934	General strike in France inaugurates the Popular Front; Luigi Pirandello wins Nobel Prize; Robert Graves's *I, Claudius* and *Claudius the God*; Frank Capra directs *It Happened One Night*
1935	Louisiana governor Huey Long assassinated; T. S. Eliot's *Murder in the Cathedral*; Hitchcock directs *The 39 Steps*
1936	Spanish Civil War begins; British King Edward VIII abdicates; Eugene O'Neill wins Nobel Prize; Margaret Mitchell's *Gone with the Wind*; Penguin Books established; *Life* magazine established; Faulkner's *Absalom, Absalom!*; John Maynard Keynes's *General Theory of Employment, Interest, and Money*; Chaplin directs *Modern Times*; death of Rudyard Kipling, Luigi Pirandello, Federico Garcia Lorca
1937	John dos Passos's *U.S.A.*; Jean-Paul Sartre's *Nausea*; Picasso paints *Guernica*; Jean Renoir directs *The Grand Illusion*; Disney releases first feature-length animated film, *Snow White and the Seven Dwarfs*; Amelia Earhart disappears; death of George Gershwin
1938	Christopher Isherwood's *Goodbye to Berlin*; Thornton Wilder's *Our Town*; Beckett's *Murphy*; Orson Welles's radio drama *The War of the Worlds*; 40-hour work-week established in the United States; death of Thomas Wolfe

1939	End of the Spanish Civil War; beginning of World War II; Joyce's *Finnegans Wake*; John Steinbeck's *The Grapes of Wrath*; Victor Fleming directs *Gone with the Wind* and *The Wizard of Oz*; John Ford directs *Stagecoach*; death of Sigmund Freud, William Butler Yeats, Ford Madox Ford
1940	Mass exodus of European artists and intellectuals to America; Leon Trotsky assassinated in Mexico; Babel secretly executed in Moscow; O'Neill's *Long Day's Journey into Night*; Richard Wright's *Native Son*; death of F. Scott Fitzgerald, Paul Klee, Walter Benjamin

Glossary

Alienation: A condition of estrangement or awareness of being cut off—from other people, from one's own sense of self, or from one's work. The term originates in philosophy and theology, but its now-central political, sociological, and psychological meanings derive from the early writings of Karl Marx. For Marx, alienation is a result of the social relations created by capitalism, which (he argues) disconnect or alienate workers from one another and from the products of their labor. These notions were refined and developed by many later writers, including sociologists and political theorists. Freud's theories of the unconscious develop a version of the self as alienated or cut off from the unconscious. Modernist art and literature are often said to describe modernity itself as an experience of alienation. In D. H. Lawrence and Kafka, humans are alienated not only from society's operations and systems, but also from their own bodies.

Center of consciousness: A term devised by the novelist Henry James, describing his alternative to the omniscient (all-knowing) perspective of most 18th- and 19th-century fiction. By restricting what is seen and known to the perspective of a single character, a single center of awareness, James argued, the story would become more plausible and vivid. His method did not require that the center of consciousness be limited to the verbal resources of the character, however, and James deploys the full range of his linguistic powers to describe the thoughts, responses, and experiences of his characters. The method is sometimes called *limited omniscience.*

Cubism: The movement in Modern art invented by Picasso and Braque in 1907. Cubism rejected the realistic representation and commitment to perspective that had defined Western painting since the Renaissance. Inspired by Cézanne's tendency toward abstraction and by African art, the Cubists substituted abstract arrangements for natural shapes. Representational elements remain in Cubist paintings, but they are flattened or distorted or arranged in unrealistic configurations. In its later phase, sometimes called *Synthetic Cubism*, its practitioners deployed fewer forms in a single painting, introduced brighter colors, and sometimes included collage elements.

Drama of the telling: Term used in these lectures to describe a signature feature of many modernist novels: their self-reflexive impulse to turn the narrator's problems in telling the story into an essentially separate story that displaces or competes with the conventional narrative.

Expressionism: In its broadest usage, this term refers to vivid distortion or expressive emphasis in works of art in many eras, cultures, and mediums. In this sense, Dostoyevsky's novels and van Gogh's paintings are Expressionist or, at least, contain Expressionist elements. In a narrower, historical sense, Expressionism is used to define the whole modern movement of the arts in Germany and Austro-Hungary in the first part of the 20th century. The movement is characterized by distortion, fragmentation, and violent emotion and by an interest in morbid psychological states. It is represented in painting (Beckmann), music (early Schonberg), literature (Kafka), and such films as Robert Weine's *The Cabinet of Dr. Caligari* (1920). The American movies called *film noir* are partly a late (and commercialized) version of Expressionism.

Impressionism: The school or movement in painting that began in France in the 1860s and was named after a Monet canvas titled *Impression: soleil levant*. This painting was displayed in 1874 at the first exhibition of a group of painters that also included Degas, Pissarro, Sisley, Renoir, and Cézanne. Emphasizing the fluidity and transience of our experience of the visible world, the movement aimed to represent the partly subjective aspect of seeing and knowing. Impressionism influenced every significant later development in 20th-century visual art. The group's last exhibition in 1886 saw the debut of Seurat and what came to be called *Neoimpressionism*. Through the example of van Gogh, the movement influenced Expressionism; in the work of Cézanne, it inspired Cubism; in the later paintings of Monet, Impressionism itself moved toward forms of abstraction that exerted a central influence on the Abstract Expressionism of the 1950s. The movement's impact on Modern literature was also profound. Joseph Conrad called himself an Impressionist; Virginia Woolf was steeped in the work and aesthetic principles of Cézanne.

Kafka-esque: An adjective applied to circumstances that recall Franz Kafka's fiction, especially *The Trial* (1925) and *The Castle* (1926), involving a nightmarish sense of futility and helplessness at the operations of powerful systems or bureaucracies that seem to obey a logic beyond human understanding.

Modernism: Broadly, the increasingly experimental movement in the arts and literature in the late 19th and the first third of the 20th century in Europe, England, and the United States. Rooted in French Impressionist painting and in the skeptical tendencies of 19th-century science and philosophy, Modernism came increasingly to define itself in opposition to the values and organizing structures of the past. Among its chief features, in both art and literature, were the following: an aesthetic self-consciousness that dramatized the limitations of artistic representation and an awareness of the materials and processes of art itself; a distrust of temporal structures and a desire to represent simultaneity or to interrupt or reorder conventional chronological sequences; an intense awareness of paradox, ambiguity, and uncertainty; a defining recognition of human subjectivity and of forces and energies that are nonverbal, irrational, and beneath conscious awareness; and a weakening of the belief in human agency and in the grand narratives and coherent ethical systems of the past. "Let us not think," wrote Virginia Woolf, "that life exists more fully in what is commonly thought big than in what is commonly thought small."

Novella: A narrative too long to be called a story and too short to be called a novel. One practical rule of thumb is widely accepted: if a story is published as a single item, even if it originally appeared in a collection of other stories, it may be promoted to novella. Some scholars see the novella, or novelette, as originating in segments or chapters in medieval story compilations. Many of the defining texts of literary Modernism are novellas, including works by Dostoevsky (in English, a "modern" writer), Henry James, Conrad, Kafka, and Thomas Mann.

Omniscient narrator: Term describing the all-knowing, God-like voice in 18th- and 19th-century novels that can move at will into the thoughts and feelings of all characters and possesses a total knowledge of their past and future. In the ancient epics, the omniscient poet narrates the thoughts and actions of the gods themselves.

Postimpressionism: A term devised by the British art critic Roger Fry to introduce an exhibition of French painting in London in 1910 and subsequently used by some art historians to designate a second stage or phase of Impressionism. All the painters represented in the exhibition were second-generation Impressionists, whose work embodied an abstracting or distorting tendency that pressed beyond the subjective realism of the first Impressionists. The show included work by Cézanne, Manet, van Gogh, Gaugin, Matisse, and Picasso.

Postmodernism: A paradoxical and perhaps unhelpful term widely used in the past 20 years to identify forms of literature, art, and architecture in the period after 1950. These works are seen as continuing the iconoclastic tendencies of Modernism but rejecting the "grand narratives" and the alleged coherence of Modernist art. Postmodern works are said to be more intensely committed to eclecticism, parody, quotation, and self-referentiality than their Modernist ancestors. The tendency—it cannot be called a movement or a school—has been associated with the rise of Poststructuralist theory and the assault on the Humanist canon of the last decades of the 20th century in England and America.

Romanticism: A vast international movement or cultural tendency that arose in the late 18th century and the beginning of the 19th century in Western Europe, Russia, England, and the United States. Romanticism was, in part, a philosophic and artistic reaction against the perceived mechanism and excessive rationalism of the Enlightenment and a response to the disruptions of belief and social organization that followed three revolutions—the French and American Revolutions and the industrial revolution. The movement has something of the same relation to the 19th century as Modernism to the 20th century. Its key elements also resemble and, in some sense, anticipate aspects of Modernism. Romanticism had a powerful psychological and subjective emphasis and, in some forms, valued the fantastic and "gothic" over more orderly, inherited art forms. It celebrated literary experiments, a poetry of feeling, and ordinary life. It involved as well a valuing of nature, instinct, and imagination against what were seen as the coercions of a mechanical reason. Romanticism had a decisive impact in England, as embodied in the poetry of Blake, Wordsworth, Coleridge, Keats, Shelley, and Byron.

Stream of consciousness: A term coined by William James, brother of Henry, and appropriated by literary scholars to identify a strategy of narration that aims to represent mental processes and the flow of conscious awareness. Often used interchangeably with the term *interior monologue*. But it is useful to understand the former term to designate all the strategies used to dramatize thought and the inner life and to allow the latter to define a more radical technique which attempts, without authorial explanation or intervention, to reproduce the random flow of memories, thoughts, half-conscious associations, and impinging outer events as they mingle and pass through the mind. In interior monologue, the level of understanding and the diction must belong entirely to the character, while in some forms of stream of consciousness (as in Virginia Woolf, for example) although the feelings and thoughts must be unique to the character, the verbal resources deployed to capture them may belong to the author herself.

Surrealism: Literally meaning "beyond realism," the term designates a general tendency in art and literature toward non-realistic distortion and the introduction of elements that contradict or ignore "reality." Surrealism was also a specific movement, founded in Paris by André Breton's *Manifesto of Surrealism* (1924), and represented by such artists as Salvador Dali and René Magritte. Inspired by 19th-century French poets and by Freud's notions of the unconscious, the movement sought to free the imagination from the constraints of rationality and consciousness.

Biographical Notes

Babel, Isaac (1894–1940). The son of a dealer in agricultural machinery, Babel was born in Odessa, whose Jewish ghetto, the Moldavanka, would be memorialized in his stories of flamboyant gangsters and their admiring narrators. Defying the restrictions of the czarist regime on Jewish travel and residence, Babel moved in 1915 to Petrograd, where his early stories were published by the eminent writer Maxim Gorky, who remained a patron and supporter of Babel until his death in 1936. During the early years of the revolution, Babel served briefly as a translator for the newly organized Cheka (secret police), wrote sketches for an anti-Leninist newspaper run by Gorky, and eventually became a war correspondent riding with a legendary Cossack cavalry regiment in the abortive Polish campaign. These experiences formed the basis of his collection of stories *Red Cavalry* (1926). When the first excerpts from this volume and his *Odessa Stories* were published in Moscow in 1923, Babel became instantly famous. Babel summarized his aesthetic of "precision and brevity" in one story written late in his career: "No iron can pierce a human heart as icily as a period in the right place." But the spare ambiguity of Babel's writing troubled the commissars, and as Stalinist conformity became the norm, Babel was attacked for his "formalism" and "estheticism." He was arrested in 1939, secretly tried on false charges of espionage, and executed in January 1940.

Beckmann, Max (1884–1950). This German Expressionist painter was influenced by Impressionism and by the Norwegian Edvard Munch (*The Scream*), but the decisive source of Beckmann's morbid, disturbing art was his experience as a medic in World War I. The maimed bodies and corpses he saw and touched during the war haunt some of his most famous paintings, including *The Descent from the Cross* (1917) and *The Night* (1918–1919), described by one scholar as "a scene of nightmarish sadism" that expressed Beckmann's "pessimism over man's bestiality." He is best known for a series of enigmatic allegorical triptychs, sometimes containing sexually explicit or sadistic imagery, in which sharply defined figures are compressed in constricting spaces. His self-portraits are also memorable, usually displaying the artist in formal attire whose elegance is disturbed or contradicted by the vacant sadness in his face and eyes.

Cézanne, Paul (1839–1906). French painter, one of the founding figures of Modernist art. Cézanne's early work belongs to the Impressionism of Renoir and Monet but moves beyond their subjective realism toward a more abstract and self-reflexive art in which geometric shapes—cones, cylinders, and spheres—partly represent or embody natural forms. In his later works, he abjured realistic perspective, creating an illusion of depth with overlapping planes. He invented a method of modeling three-dimensional forms by using patches of color in which warm colors represent planes advancing toward the viewer and cool colors signal receding planes. These tendencies in Cézanne anticipate Cubism and other later Modernist schools. Virginia Woolf's lyrical accounts of Lily Briscoe's painting in *To the Lighthouse* are widely thought to reflect her experience of Cézanne's paintings. Picasso called Cézanne "the father of us all," and he continues to influence both writers and painters. "I miss, in much contemporary writing," said the American novelist John Updike, a "sense of self-qualification, the kind of timid reverence for what exists that Cézanne shows when he grapples for the shape and shade of a fruit through a mist of delicate stabs."

Conrad, Joseph (1857–1924). Born Jósef Teodor Konrad Nalecz Korzeniowski in the Polish Ukraine, Conrad was orphaned at the age of 12. His father Apollo was a Polish nationalist who had been exiled (with his family) for revolutionary activities by the czarist governors of Russian-occupied Poland. At his death he was given a grand public funeral by the citizens of Krakow; the young Conrad walked at the head of the procession. After six years in the home of his maternal uncle, Conrad migrated to Marseilles, where he embarked on his career as a sailor. Changing his name to Joseph Conrad, he became a British subject in 1886 and a master mariner, retiring from the sea in 1894 and settling in England to become a writer. English was Conrad's third language, after his native Polish and the French he had learned as a child, but all his fiction was written in English. Many critics have observed that there is a notable foreign or French tonality in Conrad's prose—a somber, flowing energy that may derive from his unique relation to the English language. The tragic experiences of his family and his early life are also said to mark his fiction, which is haunted by themes of betrayal and political oppression. His first novel, *Almayer's Folly* (1895), shows the influence of the French novelist Flaubert. But Conrad discovers his distinctive voice in his early masterpiece, *The Nigger of the*

Narcissus (1897), described by many as the greatest novel of the sea after Melville's *Moby Dick*. Conrad drew on his 20 years as a sailor in much of his fiction, including *Heart of Darkness* (first published in *Blackwood's Magazine*, 1899), *Lord Jim* (1900), *The Secret Sharer* (1912), and *The Shadow-Line* (1917). His major political novels include *Nostromo* (1904); *The Secret Agent* (1907), often said to be the first spy novel in the English language; and *Under Western Eyes* (1911). Although he struggled financially for nearly the first 20 years of his writing life, he achieved fame and modest financial security in his last years and was buried in Canterbury Cathedral under his original (and unfortunately misspelled) Polish name. These lines from the English poet Edmund Spenser are inscribed on his tombstone: "Sleep after toyl, port after stormie seas,/Ease after warre, death after life, does greatly please."

Einstein, Albert (1879–1955). German-born theoretical physicist whose special theory of relativity revolutionized our understanding of the physical universe. He was awarded the Nobel Prize for Physics in 1921 for his early work on quantum theory. Visiting the United States when Hitler came to power, Einstein decided not to return to his native country and remained here for the rest of his life. In his last decade, he was a leading figure in the campaign against the proliferation of nuclear weapons.

Faulkner, William (1897–1962). The greatest American Modernist was born and raised in Mississippi, where he also spent much of his adult life. A high school dropout, he held a series of jobs, including postmaster and night watchman (which allowed him to write on the job), and briefly attended classes at the University of Mississippi. His first publication was a volume of poems, *The Marble Faun* (1924), followed by two novels—*Soldier's Pay* (1926) and *Mosquitoes* (1927)—widely regarded as apprentice work. He found his voice and his enduring subject matter in 1929 when he published two novels, *Sartoris* and *The Sound and the Fury*, his first masterpiece. These books are the first in his series of novels and stories about Yoknapatawpha County, an extended historical chronicle of the American Deep South from the time of the white incursions into Indian lands through the Civil War and into the 20th century. In the following 12 years, Faulkner produced a second book of poems, three volumes of stories, and eight novels, including *As I Lay Dying* (1930), *Light in August* (1932), and *Absalom, Absalom!*

(1936). In these complex masterworks, Faulkner experiments with point of view, telling his stories in the voices of a range of characters, disrupting chronological sequence, and sometimes deploying a surging, ornately adjectival prose full of extended parentheses, long passages in italics, and sentences that run over several pages. Characters and situations that appear in one story or novel return in others, as Faulkner's imagination overflows the conventional boundaries of chapter, story, or single novel. His overarching fable is an essentially tragic account of the despoliation of the wilderness, the decay of the Old South, and the abuse and exploitation of blacks and poor whites whose capacity to suffer and endure incites their creator (and the reader) to woe and wonder. Despite their brilliance, Faulkner's great works of the 1930s and early 1940s sold poorly, and his books were mostly out of print in 1946, when Malcolm Cowley's *The Portable Faulkner*, using stories and excerpts from the novels, organized the Yoknapatawpha material chronologically (and far more simply than Faulkner's tangled narratives themselves). This generated new readers and critical acclaim. In 1949, Faulkner won the Nobel Prize for Literature. In the next year, his *Collected Stories* (1950) won the National Book Award.

Ford, Ford Madox (1873–1939). Originally named Ford Hermann Hueffer, this writer was the son of a prominent music critic who wrote for *The Times* of London and was the grandson of Ford Madox Brown, a widely admired Pre-Raphaelite painter. Growing up in artistic circles, Ford published fairy stories and a novel before he was 19. He was introduced to Joseph Conrad in 1898 and collaborated with the older writer on three works of fiction during this early phase of Conrad's career. The value of this partnership for Conrad has perhaps been underestimated, for in this period Conrad developed his strategy of using first-person narrators who forgo or cannot follow a chronological ordering of events and whose struggles to understand and recount the story become a crucial aspect of the narrative. Ford describes this "impressionism," as he calls it, in an illuminating memoir titled *Joseph Conrad: A Personal Remembrance* (1924). This "impressionism" was Ford's method, as well, expressed most powerfully in *The Good Soldier* (1915) and in a tetralogy of novels titled *Parade's End* (1924–1928), about the transformation of English culture by World War I. These books contain memorable scenes of trench warfare that draw on Ford's own experiences of combat in France. Ford's contributions to Modern literature as an editor and supporter of other writers were notable. As editor of the *English Review* in 1908–1909, he discovered and was the first to publish work by D. H. Lawrence, and he also published work by Thomas Hardy, Henry James, Conrad and H.G. Wells. A generation later, editing *The Transatlantic Review* in the 1920s, he published work by Hemingway, Joyce, E. E. Cummings, Ezra Pound, and Gertrude Stein. Ford produced some 80 books during his career—32 novels, along with travel books, memoirs, and literary criticism. He taught at Olivet College in Michigan in his last years, where he was fond of describing himself as "an old man mad about writing."

Freud, Sigmund (1856–1939). Austrian neurologist and psychologist, founder of psychoanalysis. Beginning with studies of hysteria in Vienna at the end of the 19th century, Freud developed theories of personality and the mind that have profoundly influenced modern ideas of the self. His conceptions of the mind as divided against itself, driven or controlled by unconscious impulses and needs; of the centrality of sexuality in human development; and of displacement and transference, in which behavior expresses desires unknown to the conscious mind, have been immensely influential. His claims for the scientific status of psychoanalysis have been widely and powerfully disputed, but his theories have been especially fruitful in literary and cultural study. His emphasis on dream analysis and on the technique of free association deeply influenced Modernist writers and painters in Europe, England, and the United States.

Homer. The name given to the supposed author of the two founding epic poems of Western civilization, the *Iliad* and the *Odyssey*. Nothing is known about Homer, and many contemporary classicists believe it is unlikely that the same man wrote both works. It is certain that both texts are compilations or recombinations of traditional materials that had been widely known and performed or recited for centuries before the advent of writing. Some theories hold that the poems were created or organized in something like the form in which we know them during the 8th century B.C.E. Homer occupied a nearly sacred position in ancient Greece, where the poems were performed regularly at religious festivals and provided subjects for poets and dramatists. The *Iliad* tells the story of the 10-year Trojan War, in which a Greek army led by Agamemnon lays siege to the city of Troy and ultimately conquers it. The *Odyssey* is a kind of sequel, in which one of the primary Greek heroes of the *Iliad*, Odysseus, known for his wiles, endures a 10-year trial by nymph and monster before returning to his home, the Greek island of Ithaca. Joyce's *Ulysses* is a modern retelling of the *Odyssey*, using the Latin name for Odysseus.

James, Henry (1843–1916). A prolific novelist and story writer whose representations of the inner life of his characters anticipated and deeply influenced Modern fiction, James was born in New York City, the second son of a prominent lecturer on religion and social theory and the younger brother of William James, the distinguished psychologist and philosopher who coined the term *stream of consciousness*. The James children were taken to Europe as infants and were home-schooled by tutors. Returning to Europe in their teens with their family, the brothers became adept at languages and comfortable with foreign cultures. These experiences formed the ground for much of Henry's fiction, whose dominant theme is the encounter or contrast between American vitality and innocence and the more worldly, aristocratic civilization of Europe. James settled permanently in Europe in 1875 and became an English citizen in the year before his death as a gesture of support for his adopted country at the outset of World War I. In a career spanning more than 50 years, James produced 20 novels and more than 100 stories and novellas, as well as plays, distinguished literary criticism, and travel writing. His major works include *Daisy Miller* and *The Europeans* (both 1878), *The Portrait of a Lady* (1881), *The Spoils of Poynton* (1897), *The Turn of the Screw* (1898), *The Wings of the Dove* (1902), and *The Ambassadors* (1903). In a series of prefaces to a collected edition of his fiction, James articulated his theory of the novel, which rejected the omniscient perspective of such 19th-century masters as Tolstoy in favor of "the muted majesty" of a point of view restricted to a primary character's "center of consciousness." These principles and James's practice as a novelist were much admired and imitated by later writers. Conrad, for example, called James his "*cher maître*."

Joyce, James (1882–1941). The most innovative and influential novelist of the 20th century, Joyce was born and educated in Dublin, Ireland, graduating from University College, Dublin, in 1903. He renounced his faith at the age of 16 and remained a critic of the Irish Catholic Church throughout his life. On June 16, 1904, he took his first walk with Nora Barnacle, the woman with whom he would spend his life; this event was commemorated as "Bloomsday," the single day on which the action of his masterpiece *Ulysses* unfolds. Later that year, Joyce and Nora left Ireland, beginning a self-imposed exile that would last, with the exception of two brief visits in 1909 and 1912, for the rest of his life. The couple settled in Trieste, now part of Italy, where Joyce taught English at the Berlitz school and where they would live for most of the next decade. Here, he met the Italian novelist-to-be Italo Svevo, whose writing he encouraged and who became a model for Leopold Bloom, the protagonist of *Ulysses*. Joyce and Nora spent the war years in Zurich, then settled in Paris in 1920, their home for the next two decades. Joyce's first publication was a negligible collection of poems, *Chamber Music* (1907). But his real energies had for some time been devoted to fiction. His collection of linked stories, *Dubliners*, appeared in 1914 after eight years of delay owing to his publisher's fear that the book would face charges of libel or obscenity.

Joyce's autobiographical novel, *A Portrait of the Artist as a Young Man*, appeared in serial form in *The Egoist* in 1914–1915 and in book form from an American publisher in 1916. Despite powerful support from such figures as Ezra Pound, W. B. Yeats, and H. G. Wells, the novel was rejected by every British publisher to whom Joyce submitted it. A similar fate awaited *Ulysses*, Joyce's complex retelling of Homer's *Odyssey* as the story of one man's ordinary experiences over the course of a single day. Chapters of the book were published in serial form in an American magazine beginning in 1918, but three separate issues were seized by authorities on the grounds of obscenity. Neither British nor American publishers were willing to publish the already-famous (and allegedly scandalous) manuscript. But Joyce's benefactress and editor of the *Egoist*, Harriett Shaw Weaver, arranged for publication in Paris in 1922. Immediately recognized in literary circles as a masterpiece, *Ulysses* existed as an underground book, banned in England and in the United States, for more than a decade. In a landmark trial in New York City in 1933 (affirmed in 1934), the novel was finally judged not to be obscene and was published legally. A British edition appeared in 1937. Joyce's final novel, *Finnegans Wake*, described by the author as his "book of the night" as *Ulysses* was his book of waking life, appeared in 1939. Perhaps Joyce's chief innovation—and a source of his conflicts with the censors—was his development of the interior monologue, in which a character's random thoughts and associations are presented without editorial intervention or explanation. This strategy makes extreme demands on readers, but it also creates characters whose inner lives and moral natures are seen with unsurpassed intimacy and fullness.

Kafka, Franz (1883–1924). This tormented genius, whose name has become synonymous with the defining anxieties of modern life, was born and raised in Prague in a German-speaking Jewish family. His religious training was minimal, although in his thirties, to the chagrin of his assimilated father, he became deeply engaged with Jewish theater and Jewish mystical traditions. Always a superior student, Kafka received a doctorate in law from the University of Prague in 1906 and worked for a year for a private insurance company before accepting employment with the partly state-run Worker's Accident Insurance Institute for the Kingdom of Bohemia in Prague. He spent 13 years at the institute, winning cases that improved working conditions for miners and other workers and attaining a position of significant authority. All his life, Kafka suffered from migraine headaches and other ailments associated with stress and depression. In 1917 he contracted tuberculosis, the illness that led to his death seven years later. Kafka published very little during his lifetime, though he worked obsessively on his fiction and often wrote through the night before going off to what he called his "bread-work" in the insurance world. The novels that established his fame—*The Trial* (1925), *The Castle* (1926), and *Amerika* (1927)—were published posthumously by his lifelong friend Max Brod, who ignored Kafka's instructions to destroy his unpublished work. These books, along with the early novella *The Metamorphosis* (1915) and a handful of other stories, dramatize a surreal or dreamlike world that is also strangely ordinary, in which baffled, anxious protagonists struggle against enigmatic circumstances for a justice or redemption that never arrives.

Kipling, Rudyard (1865–1936). The son of John Lockwood Kipling, principal of the Jeejeebhoy School of Art and Industry and author and illustrator of *Beast and Man in India* (1891), Kipling was born in Bombay. He was taken to England with his sister at age six and spent five years, separated from his parents, with a Portsmouth family who took in the children of Anglo-Indians. This experience generated two of Kipling's most powerful works, the bitter story

"Baa Baa, Black Sheep" (1888) and the novel *The Light That Failed* (1890). He spent the years 1878–1882 at a military school in Devon, and then he returned to India to work as a journalist. His early stories and poems were published first in newspapers and later collected in such books as *Departmental Ditties* (1886), *Plain Tales from the Hills* (1888)—which included *The Man Who Would Be King*—and *Wee Willie Winkie* (1890). He came to London in 1889 and achieved almost instant fame, in part because some of his poetry in the colloquial voices of British soldiers serving in India was published in a magazine in the United Kingdom. These poems, including "Gunga Din" and "Danny Deever," were collected as *Barrack-Room Ballads* (1892). Other works include *Kim* (1901), often said to be his best novel, and his tales for children, *The Jungle Book* (1894), *The Second Jungle Book* (1895), and *Just So Stories* (1902). In 1907, Kipling became the first British writer to win the Nobel Prize for Literature.

Lawrence, D(avid) H(erbert) (1885–1930). Lawrence was born in the English mining town of Eastwood, Nottinghamshire to a refined and literary former schoolteacher and a hard-drinking, nearly illiterate coal miner. The marriage was an unhappy one; the parents quarreled harshly over their sickly, gifted son; Lawrence's mother was eager to keep him out of the mines and encourage his desire for education. He did avoid the mines, becoming a pupil-teacher, then winning a scholarship to Nottingham University, where he earned a teacher's credentials. He taught for two years in an elementary school while writing poems and stories but was told by doctors to give up teaching after a serious illness, perhaps connected to the death of his mother. These early experiences are re-created in Lawrence's searing novel, *Sons and Lovers* (1913). In 1912, he met Frieda Weekley, the wife of one of his former professors, and eloped with her to Germany, beginning a turbulent relationship that would last until his death. The couple spent much of their time traveling, not only on the continent but also on extended visits to Asia, Australia, and North America. Despite bouts of illness and a constant lack of money, Lawrence wrote prodigiously during these years, producing travel books, poetry, works on psychology, and a series of stories and novels, including *The Rainbow* (1915); its sequel, *Women in Love* (written 1916, published 1920); *Aaron's Rod* (1922); *Kangaroo* (1923); and *The Plumed Serpent* (1926). The sexual candor of Lawrence's fiction incited outrage and censorship. *The

Rainbow and *Women in Love* were suppressed in England as obscene. Lawrence's most explicit treatment of sex occurs in his final novel, *Lady Chatterley's Lover*, privately published in Italy in 1928 but banned in England and the United States for more than 30 years.

Monet, Claude (1840–1926). French painter, one of the founding Impressionists, and the greatest in a group of great artists that included Renoir, Sisley, Pissarro, and others. The term *Impressionism* derives from his 1872 canvas titled *Impression: soleil levant* (*Impression: Sunrise*). His signature works are series-paintings of the same subject: haystacks, the Paris train station, the houses of Parliament in London as seen across the Thames, and the Gothic cathedral in Rouen. These series are simultaneously exact representations of their subjects as they appear in the changing light and weather *and* self-conscious meditations on the nature and materials of art. In Monet's mature paintings, "reality" is understood to be unstable, always in process, made and unmade not only by changes in light and climate but also by the partly subjective perspective of the painter or viewer. Monet endured neglect and extreme poverty for many years, but the modest success of an exhibition of his paintings in 1883 allowed him to buy a small property in Giverny, where he lived for the rest of his life. The lily pond on this property became one of his great subjects, represented in scores of paintings that are displayed today in museums around the world.

Muybridge, Eadweard (1830–1904). English photographer whose studies of humans and animals in motion are now recognized as a forerunner of the movies. Using a series of cameras with special shutters, Muybridge was able to capture minute gradations of movement. He invented a special apparatus (a precursor of the movie projector) to display his images, which were published under the title *Animal Locomotion* in 1887. This portfolio was a primary inspiration for Duchamp's revolutionary painting *Nude Descending a Staircase No. 2* (1912).

Nabokov, Vladimir (1899–1977). The son of V. D. Nabokov, a Russian aristocrat and lawyer who was a leader of the non-socialist Constitutional Democratic Party in pre-revolutionary Russia. V. D. Nabokov drafted the Romanov abdication manifesto and served in the provisional government of 1917. Following the Bolshevik coup d'état in 1919, he went into exile with his family, settling in Berlin. He was assassinated in 1922 by czarist extremists at a political meeting as he tried to shield another speaker. His son would dramatize this absurd, mistaken assassination several times in his fiction, notably in the climactic event of *Pale Fire* (1962). The themes of exile and displacement are central to nearly all of Nabokov's novels, and his admiration for his father and his unappeasable sense of loss over the world of his childhood are the heart of *Speak, Memory* (1967; originally published in 1951 as *Conclusive Evidence*), a compelling memoir some have called his finest and subtlest book. Nabokov studied French and Russian literature at Trinity College, Cambridge, and then lived in Berlin and Paris until 1937, publishing poetry and fiction in Russian under the pseudonym V. Sirin. He emigrated to the United States in 1940, teaching first at Wellesley College, then at Cornell. His first nine novels were written in Russian and later translated into English with Nabokov's collaboration. But beginning with *The Real Life of Sebastian Knight* (1941), Nabokov's fiction was written in a brilliantly inventive English. The eight novels written in English are commonly recognized as his best work, particularly *Lolita* (1955), the scandalous bestseller narrated by a mad verbal genius who is also a pedophile; *Pnin* (1957), perhaps this cool novelist's warmest book, about a Russian émigré teacher of languages befuddled by the speed and materialism of American life; and *Pale Fire*, a unique parody-novel disguised as an edition of a narrative poem. The financial success of *Lolita* permitted Nabokov to give up teaching and move to Montreux, Switzerland in 1959, where he lived with his wife, Vera, until his death.

Picasso, Pablo (1881–1973). The most famous and influential visual artist of the 20th century. Born in Spain, Picasso worked primarily in France. Best known for his painting, he also produced prolifically in many other fields, including sculpture, ceramics, engraving and other forms of printmaking, and book illustration. A child prodigy (by the age of 10, his drawings were recognized as marks of genius), Picasso had his first exhibition at 13 and won honorable mention at 18 for a painting shown in Madrid at the Fine Arts Exhibition. Shortly after this, however, the young genius rebelled against both the traditional teaching he encountered at the Royal Academy of San Fernando and his family's expectations for his success as an academic painter. Following what have come to be called his "blue" and "rose" periods

(1901–1906)—during which these colors dominated his often elongated or distorted, but still realistic, figures—Picasso embarked on a career of unprecedented inventiveness and productivity. His seminal work, *Les Demoiselles d'Avignon* (1907), drew inspiration both from Cézanne and from African art. The painting shocked traditionalists with its mask-like treatment of faces, its bold distortions, and its rejection of realistic perspective. (The title of the painting refers to Avignon Street in Barcelona, known for its brothels, so the implication that the *demoiselles* are prostitutes was a further source of scandal.) The *Demoiselles* is often identified as the first work of Cubism, the movement Picasso and his friend Braque invented in the period 1907–1914. Although he never became an official member of the Surrealists, Picasso's work influenced that school and was, in turn, enriched by it, in particular by its use of dreamlike imagery and erotic themes. Perhaps his most widely known painting is the powerful antiwar allegory *Guernica* (1937), whose title names the Basque city in northern Spain that was bombed by German planes during the Spanish Civil War.

Van Gogh, Vincent Willem (1853–1890). Dutch painter. All Van Gogh's work was created in the last decade of his life, and most of that within his astonishingly productive (and tortured) last four years. The son of a Protestant minister, he endured an unsettled and solitary early adulthood, working briefly for a firm of art dealers (of which his uncle was a partner), as a bookseller and lay preacher, and finally, as an evangelical missionary among the poor in a mining region in Belgium. Dismissed from this post after giving away all his goods in a gesture of solidarity with his impoverished clients, Van Gogh began to draw and discovered his true vocation. From 1880–1885, he worked mainly in the Netherlands and Belgium, studying with other painters and enrolling as an art student in the Antwerp Academy. In 1886, he joined his brother Theo, an art dealer living in Paris, where he made contact with a group of Postimpressionists, including Toulouse-Lautrec, Gauguin, Pissarro, and Seurat. Their work and that of other French painters opened Van Gogh to new uses of color and line and almost immediately led him to develop his distinctive style of swirling brush strokes and vivid colors. Van Gogh's mature style was spontaneous; he worked with tremendous speed, sometimes squeezing paints from their tubes directly onto the canvas. In letters to Theo, Van Gogh explained that the vivid intensity of his colors aimed to express an "inner fire" of emotion and authenticity not visible in the external world. Van Gogh suffered from mental illness that became more intense in his final years. After an argument with Gauguin, he mutilated himself, slicing off a portion of his right ear. This melancholy event is commemorated in one of his greatest paintings (profoundly lucid in its depiction of the artist's empty gaze), *Self-Portrait with Bandaged Ear* (1889). Van Gogh committed suicide in July 1890, having produced more than 800 oil paintings and 700 drawings. He sold only one painting during his lifetime and was completely unknown at the time of his death. A century later, according to one source, he had become the most recognized painter of all time.

Woolf, Virginia (1882–1941). A frail and nervous but brilliant child, Virginia Woolf was the daughter of Leslie Stephen, an eminent intellectual, author, and founding editor of Britain's *Dictionary of National Biography*. She grew up surrounded by gifted siblings and half-siblings: her sister, two brothers, and four older children from her parents' previous marriages. The Stephen home in Kensington teemed with distinguished visitors, including such great writers as Henry James and George Eliot, and its large library provided Virginia's higher education (although her brothers, of course, attended university). The death of her mother in 1895 triggered the first in a series of mental breakdowns that were to persist throughout her life, culminating in her suicide by drowning on the eve of World War II. But the story of her life is also one of astonishing intellectual vitality and productivity. Recognized today as one of most innovative of Modern novelists, Woolf was also the finest literary critic of her age, one of the founding voices of modern feminism, and a dazzling letter-writer and diarist whose writing in these modes (now collected in 11 volumes) constitutes a cultural history of intellectual Britain between the wars. Two of her eight novels—*Mrs. Dalloway* (1925) and *To the Lighthouse* (1927)—are among the great fictions of the century, composed in a lyrical stream-of-consciousness style that dramatizes human interiority with unique subtlety and authority. Her monograph, *A Room of One's Own* (1929), begun as a lecture to female students at Cambridge, articulates a trenchant and courageous critique of patriarchy. Her literary essays, first gathered in *The Common Reader* (1925) and *The Common Reader: Second Series* (1932), include thoughtful readings of her literary ancestors and contemporaries as well as influential general pieces on the morally revolutionary nature of Modern literature.

Bibliography

Essential Reading:

Babel, Isaac. *The Collected Stories.* Edited by Nathalie Babel, translated by Peter Constantine. New York and London: Norton, 2002. Eloquent introduction by Cynthia Ozick.

Conrad, Joseph. *Heart of Darkness.* Edited by Paul B. Armstrong. New York: Norton, 2006. This fourth edition of the Norton Critical edition contains historical background material and a sampling of interpretative criticism.

———. *The Shadow-Line: A Confession.* Edited by Jeremy Hawthorn. Oxford and New York: Oxford University Press, 2003. Includes brief annotations and a thoughtful introduction.

Cowley, Malcolm, ed. *The Portable Faulkner.* New York: Penguin, 1978. This revised version of the 1946 collection that revived Faulkner's reputation in the United States includes "That Evening Sun" (pp. 391–410) and "Old Man" (481–581), as well as the famous map of Yoknapatawpha County, "surveyed and mapped for this volume" by the author.

Faulkner, William. *Absalom, Absalom!* New York: Random House, 1986. Corrected version of the novel, originally published in 1936.

Ford, Ford Madox. *The Good Soldier.* New York: Vintage, 1989. Mark Schorer's "An Interpretation," which introduces this edition, has been influential.

Hamilton, George Heard. *Painting and Sculpture in Europe, 1880–1940.* London: Penguin, revised edition, 1987. This volume in the distinguished series, Pelican History of Art, is beautifully written and generously illustrated.

Joyce, James. *Dubliners.* Edited by Terrence Brown. New York: Penguin, 1993. An annotated edition that includes an appendix listing characters from the stories who reappear in *Ulysses*.

———.*Ulysses.* New York: Vintage, 1961. A revised version of the first American edition published in 1934. The so-called "corrected edition," edited by Hans Walter Gabler (Vintage, 1986), is also available and is preferred by some scholars but introduces new errors and makes some dubious typographical choices.

Kafka, Franz. *The Metamorphosis.* Translated and edited by Stanley Corngold. New York: Bantam Books, 1987. Includes excerpts from key biographical documents and critical essays. The older translation by Willa and Edwin Muir, widely available in anthologies, is preferred by some scholars. One source for this translation is Kafka's *Complete Stories*, cited below in Supplementary Reading.

Kipling, Rudyard. *The Man Who Would Be King and Other Stories.* Edited by Louis L. Cornell. Oxford and New York: Oxford University Press, 1999. Contains a good selection of other Kipling stories, including his famous tale of childhood suffering, "Baa Baa, Black Sheep."

Lawrence, D. H. "The Horse Dealer's Daughter," in *The Complete Short Stories of D. H. Lawrence.* New York: Viking, 1950. Also available in many anthologies of English fiction, including pp. 199–211 of *Initiation*, edited by Thorburn, cited below in Supplementary Reading, as well as in *Short Story Masterpieces.* Robert Penn Warren and Albert Erskine, eds. New York: Random House, 1954.

Nabokov, Vladimir. *Pale Fire.* New York: Vintage International, 1989. First published in 1962.

Ransome, John Crowe. "Captain Carpenter," in *Modern Poems.* Edited by Richard Ellmann and Robert O'Clair. New York: Norton, 1989, pp. 266–267. This classic Modernist poem is available in many anthologies. This Norton Introduction contains a rich selection of poets from Whitman through the High Modern classics to contemporary figures and concludes with a valuable essay titled "Modern Poetry in English: A Brief History."

Scholes, Robert, ed. *Approaches to the Novel.* San Francisco: Chandler, 1961. This collection of classic essays by critics and novelists is out of print but available in good libraries and as a used book in many online venues. It includes Norman Friedman's important article on point of view in Modern fiction and Virginia Woolf's seminal essay, "Mr. Bennett and Mrs. Brown," first printed by the Hogarth Press as a pamphlet in 1928.

Woolf, Virginia. *To the Lighthouse.* New York: Harcourt Brace Jovanovich, 1989. Sturdy paperback with a foreword by Eudora Welty. First published in 1927.

Supplementary Reading:
General Titles

Eliot, T. S. "The Waste Land," in *Modern Poems*. Edited by Richard Ellmann and Robert O'Clair. New York: Norton, 1989, pp. 280–293. A good edition.

Ellmann, Richard, and Charles Feidelson, eds. *The Modern Tradition: Backgrounds of Modern Literature*. New York: Oxford University Press, 1965. The standard source for the theoretical backgrounds of Modernism, this collection contains excerpts from a range of Modernist artists—including Wilde, Rilke, Joyce, Picasso, Eliot, Mallarmé, and many others—as well as philosophers and cultural theorists.

Frank, Joseph. *The Widening Gyre: Crisis and Mastery in Modern Literature*. Bloomington: Indiana University Press, 1968. Includes the influential essay, "Spatial Form in Modern Literature." first published in 1945.

Freud, Sigmund. *The Future of an Illusion*. Translated by W. D. Robson-Scott. Revised and newly edited by James Strachey. Garden City, NY: Anchor Books, 1964. Translation by the one editor Freud authorized.

Homer. *The Odyssey*. Translated by Robert Fitzgerald. New York: Farrar, Straus and Giroux, 1998. This translation, first published in 1961, remains the best contemporary English version of the poem in the opinion of many scholars and teachers. The edition includes an informative introduction by the distinguished classicist D. S. Carne-Ross.

Hynes, Samuel. *The Edwardian Turn of Mind*. Princeton: Princeton University Press, 1968. Great cultural history of England's transition from the Victorian to the Modern era. Includes an incisive chapter on the 1910 Postimpressionist Exhibition in London.

Kahler, Erich. *The Inward Turn of Narrative*. Translated by Richard and Clara Winston. Princeton: Princeton University Press, 1973. Compelling study of Modern fiction's preoccupation with human subjectivity.

Kershner, R. B. *The Twentieth-Century Novel: An Introduction*. Boston and New York: Bedford Books, 1997. Useful, short handbook containing extensive bibliographies of critical and theoretical accounts of Modern fiction.

Kiely, Robert. *Beyond Egotism: The Fiction of James Joyce, Virginia Woolf, and D. H. Lawrence.* Cambridge, MA: Harvard University Press, 1980. Provocative thematic chapters with such titles as "Friendship," "Marriage," and "Actor and Audience."

Riding, Alan. "Monet's Fixation on the Rouen Cathedral," *New York Times,* 15 August 1994, pp. C9–10. Describes the 100th-year commemoration of the cathedral paintings in their source city.

Thorburn, David, ed. *Initiation: Stories and Novels on Three Themes.* New York: Harcourt Brace Jovanovich, 1976. Now out of print but available in libraries and as a used book online, this collection contains an introduction that reprises the second lecture in this series and includes, among many other Modern stories, four of the primary texts examined in this course: Lawrence, "The Horse Dealer's Daughter"; Kafka, *The Metamorphosis*; Conrad, *The Shadow-Line*; and Babel, "The Story of My Dovecot."

Van Ghent, Dorothy. *The English Novel: Form and Function.* New York: Holt, Rinehart and Winston, 1953; reprint, 1960. Includes luminous chapters on *Lord Jim, Sons and Lovers,* and *A Portrait of the Artist as a Young Man.*

Individual Authors

Babel:

Charyn, Jerome. *Savage Shorthand: The Life and Death of Isaac Babel.* New York: Random House, 2005. Controversial but plausible biography that stresses the ambiguity of Babel's connections with Soviet authority.

Freidin, Gregory, ed. *Isaac Babel's Selected Writings.* Translated by Peter Constantine. New York: Norton, scheduled for publication in May 2007. This unusual Norton Critical edition will contain extensive material on Babel as his contemporaries viewed him, as well as a significant selection of his letters, a chronology of his life, and a gathering of recent interpretive articles. Every story discussed in these lectures will be included in the volume.

Conrad:

Conrad, Joseph. *Lord Jim.* Edited by Thomas Moser. New York: Norton, 1996. Norton Critical edition.

———. *Nostromo.* New York: Modern Library, 1983. Introduction by Robert Penn Warren.

———. *The Secret Sharer*, in *Heart of Darkness and The Secret Sharer*. New York: Signet, 1983. Introduction by Albert Guerard.

Fogel, Aaron. *Coercion to Speak: Conrad's Poetics of Dialogue*. Cambridge, MA: Harvard University Press, 1985. A brilliant, undervalued account of Conrad's major fiction.

Guerard, Albert. *Conrad the Novelist*. Cambridge, MA: Harvard University Press, 1962. Definitive study of Conrad's psychological themes.

Thorburn, David. *Conrad's Romanticism*. New Haven: Yale University Press, 1974. Addresses the continuities between Romantic and Modern literature and Conrad's revision of the conventional adventure narrative.

Faulkner:

Faulkner, William. *The Sound and the Fury*. Edited by David Minter. New York: Norton, 1987. This Norton Critical edition contains significant biographical and cultural background, as well as interpretive essays. Minter is also the author of a fine biography of Faulkner.

Brooks, Cleanth. *William Faulkner: The Yoknapatawpha Country*. New Haven: Yale University Press, 1966. Still a definitive study.

Hobson, Fred, ed. *William Faulkner's Absalom, Absalom!: A Casebook*. New York: Oxford University Press, 2003. Contains essays stressing the themes of history, race, and gender.

Ford:

Ford, Ford Madox. *The Good Soldier*. Edited by Martin Stannard. New York: Norton, 1995. This Norton Critical edition includes a selection of contemporary reviews, a valuable section on Ford's "literary impressionism," and biographical and analytical essays.

———. *Parade's End*. New York: Knopf, 1961. Introduction by Robie Macauley.

Joyce:

Joyce, James. *A Portrait of the Artist as a Young Man*. New York: Viking, 1964. Joyce's autobiographical novel.

Blamires, Harry. *The New Bloomsday Book: A Guide through Joyce's Ulysses*. 3rd ed. New York: Routledge, 1996. A chapter-by-chapter plot summary. Serious readers will work through each chapter on their own before turning to this helpful resource.

Ellmann, Richard. *James Joyce*. New York: Oxford University Press, 1959; reprint 1965. One of the great literary biographies in English.

Gifford, Don and Robert J. Seidman. *Ulysses Annotated: Notes for James Joyce's Ulysses*. Berkeley: University of California Press, 1989. Revised and much enlarged from its original 1974 publication, this is an indispensable reference work.

Schwarz, Daniel R. *Reading Joyce's Ulysses*. London: Palgrave MacMillan, 2004. Reissued and updated from its 1987 publication, this is a superior general introduction to the novel, emphasizing the relationship between Joyce's formal experiments and the human story.

Kafka:

Kafka, Franz. *The Complete Stories*. Edited by Nahum N. Glatzer. New York: Schocken Books. Foreword by John Updike. Contains all of Kafka's fiction except for the three novels.

———. *The Trial*. Translated by Breon Mitchell. New York: Schocken Books, 1983. Contains a chronology of Kafka's life. Based on the new German edition of Kafka's works.

Corngold, Stanley. *Lambent Traces: Franz Kafka*. Princeton: Princeton University Press, 2004. Elegant, magisterial essays on all aspects of Kafka's art.

Kipling:

Kipling, Rudyard. *Kim*. Edited by Máire ní Fhlathúin. Ontario: Broadview Editions, 2005. Rich critical edition, especially strong on the novel's relation to British imperial history.

Wilson, Angus. *The Strange Ride of Rudyard Kipling*. New York: Viking, 1977. A rare "life and works" that is as perceptive about Kipling's fiction as it is about his life. Vivid, lively prose.

Lawrence:

Lawrence, D. H. *Lady Chatterley's Lover*. Edited by Michael Squires. New York: Penguin, 2006. Introduction by Doris Lessing.

———. *Sons and Lovers*. New York: Modern Library, 1999. Introduction by Geoff Dyer.

———. *Women in Love*. New York: Modern Library, 2002. Introduction by Joyce Carol Oates.

Moynahan, Julian. *The Deed of Life: The Novels and Tales of D. H. Lawrence*. Princeton: Princeton University Press, 1963. Insightful and jargon free.

Nabokov:

Nabokov, Vladimir. *The Annotated Lolita*. Edited by Alfred Appel, Jr. New York: McGraw Hill, 1970. Nabokov is respectfully "Kinboted" in this instructive edition of his most scandalous novel.

Alter, Robert. *Partial Magic: The Novel as a Self-Conscious Genre*. Berkeley: University of California Press, 1975. Chapter 6, "Nabokov's Game of Worlds," about *Pale Fire*, is one of the best essays ever written on this riddling author.

Boyd, Brian. *Vladimir Nabokov: The American Years*. Princeton: Princeton University Press, 1991. Second volume of the most thorough biography, covering the years of *Lolita* and *Pale Fire*.

Woolf:

Woolf, Virginia. *Mrs. Dalloway*. New York: Harcourt, 1981. Foreword by Maureen Howard.

———. *The Common Reader*. New York: Harcourt, Brace, 1953. A sampling of Woolf's best criticism, including the landmark essay "Modern Fiction" and pieces on Austen, George Eliot, Dostoyevsky, and Conrad.

———. *A Room of One's Own*. New York: Harcourt Brace, 1957. Woolf's famous brief for intellectual equality for women.

Lee, Hermione. *Virginia Woolf*. New York, Vintage, 1999. Superb biography.

Naremore, James. *The World Without a Self: Virginia Woolf and the Novel*. New Haven: Yale University Press, 1973. On Woolf's notions of consciousness and personality.

Internet Resources:

Because the Internet addresses (URLs) of many websites are often transient, users must persist when searching online for resources. Below are some useful sites, operational as of January 1, 2007:

The Modern Word. A site sponsored by the Gotham Writers' Workshop that has remained stable for some years. It describes itself as "the leading creative writing school in NYC and the United States." A lively, serious space, it includes what it terms "archives" on Beckett, Pynchon, Joyce, Kafka, and Gabriel Garcia Marquez. http://www.themodernword.com.

James Joyce. A site maintained by the Columbia University Joyce scholar Michael Seidel that includes streaming audio of many songs mentioned and performed in *Ulysses.* http://www.columbia.edu/itc/english/seidel/joyce/.

The Victorian Web. A Victorian culture site that includes material on late-19th-century texts, including works by Kipling and Conrad. http://www.victorianweb.org/.

William Faulkner on the Web. A Faulkner site containing a good biography, some analysis of texts, a glossary of characters, and key terms. http://www.mcsr.olemiss.edu/~egjbp/faulkner/faulkner.html.

Zembla. A remarkable site devoted to Nabokov, full of useful information, images, critical essays, and imaginative creative work. http://www.libraries.psu.edu/nabokov/zembla.htm.

Notes

Notes